2

FUNDAMENTALS OF ENVIRONMENTAL MANAGEMENT

FUNDAMENTALS OF ENVIRONMENTAL MANAGEMENT

Steven L. Erickson

Brian J. King

JOHN WILEY & SONS, INC.

New York / Chichester / Weinheim / Brisbane / Singapore / Toronto

This book is printed on acid-free paper. ♾

Published simultaneously in Canada.

Library of Congress Cataoging-in-Publication Data:
Erickson, Steven L.
Fundamentals of environmental management / Steven L. Erickson,
Brian J. King.
p. cm.
Includes index.
ISBN 0-471-29134-X (cloth)
1. Environmental management. 2. Environmental law—United States.
I. King Brian J. II. Title.
GE300.E75 1999
658.4'08—dc21 98-45206

Printed in the United States of America.

10 9 8 7 6 5 4 3 2 1

To the environmental professionals who devote their careers to improving and protecting our environment

CONTENTS

FOREWORD

Today's environmental professional may quite naturally be daunted by the plethora of regulations with which his or her company must comply. Federal regulations alone constitute literally thousands of pages of complicated requirements, to say nothing of the additional expectations spelled out in state and local laws. The proliferation of new regulations continues—new requirements to publicly disclose more and more about the chemicals manufactured or emitted by a facility, together with information on how and where they are used and stored. This new information quite naturally raises health concerns with the public that, in turn, must be addressed. As if this were not enough, there is a whole new collection of so-called voluntary programs, both federal and state, such as EPA's XL, Green Lights, or the SARA 33/50 program, the merits of which must be evaluated by environmental professionals.

All of these new requirements are coming at a time when industry is in the midst of massive mergers and consolidation. Many companies, even though not directly impacted by these actions, find they must downsize or rightsize to remain competitive, not only nationally but internationally as well. The resulting reduction in staff sizes means that a greater number of responsibilities have been shifted to a fewer number of professionals. This, in turn, has required the focus to shift from a cadre of environmental "specialists" (air, water, solid and hazardous wastes) to personnel who must be more general in their knowledge. Additional training in new fields is the inevitable result.

There may once have been a time when strict adherence to the legal requirements of these laws and regulations would suffice. That doesn't seem to be enough today, as the public expects companies to go beyond compliance in their environmental endeavors and to consider concepts such as "sustainability" and "life cycle analysis." The emergence of ISO standards for environmental management systems furthers these new concepts, including a requirement to evaluate the impacts of a facility's operations on the environment.

As imposing as strict compliance with existing regulations is, some companies desire to go beyond mere compliance to differentiate themselves from their competitors with their environmental programs in order to gain some market advantage. This objective creates additional challenges for the environmental staff of such companies and emphasizes the need for ever greater understanding of the environmental laws and regulations.

All of this would be more manageable if one had access to a comprehensive review of the fundamental concepts of environmental management as a guidance document. I believe that this book, *Fundamentals of Environmental Management,* fills that need. Erickson and King have managed to capture the essential elements and salient points of the most relevant environmental regulations and have summarized them in an easily understood format, which will serve as a valuable reference tool for the busy environmental professional. In this book, I believe the reader will find practical advice concerning environmental permitting and the elements of a good environmental management system. Also contained here are guidelines for dealing with environmental regulators and the public. For those companies interested in going ''beyond compliance,'' there are several useful ideas as well.

This book will be of considerable help in understanding and complying with the world of environmental regulations. The consequences of noncompliance and the associated civil and criminal penalties can be catastrophic to a company, and the resulting publicity can jepoardize the company's business. The simple fact is that a facility must earn the right to continue to operate, in the eyes of both the regulators and the public. Obtaining and maintaining valid environmental permits constitutes a critical component of the ''operating license'' for a facility and allows the facility to concentrate its efforts on conducting a profitable operation. In short, maintaining a good environmental compliance program is just good business.

Vice President, Environmental Quality
Champion International
Stamford, Connecticut

ED CLEM

PREFACE

The management of industrial environmental affairs has become increasingly complex owing to the expanding maze of environmental laws and regulations and the growing public expectations regarding environmental protection. In a review of the available literature, we learned that environmental management books typically fall into one of two categories. First, there are a multitude of environmental legal books written by attorneys for attorneys, which provide exhaustive detail regarding the legal requirements of the environmental laws but little or no advice about how those laws are applied in practice. There also are several books that examine the business aspects of environmental management, but with no dicussion of the basic legal framework. In our book we have combined the management and legal aspects of environmental management in a single volume that offers industrial environmental managers a broad overview of the fundamental elements of environmental management and a summary of the environmental laws important to a facility's program.

This book offers a comprehensive review of the environmental laws and regulations, with a quick reference to the pertinent citations. The chapters summarizing the environmental laws discuss regulatory compliance and provide practical advice on how to implement these laws and obtain environmental permits and regulatory approvals. Although these summaries are not a substitute for a detailed review of the statutes and regulations, they offer a good starting point for any study of the regulatory requirements. Since environmental laws are in a constant state of flux, we caution the reader against relying completely on the forms and lists we have included in our book. Always check with the authorized agency to be sure you have the latest version of any regulatory materials.

Several chapters are devoted to outlining the basic elements of environmental management programs and assessing the environmental impacts associated with industrial development. We also decided to look beyond basic environmental compliance and examine programs such as ISO 14000 that will likely form the basis of the next wave of environmental regulation. We emphasize the importance of maintaining good relationships with regulatory agencies and public stakeholder groups and provide a detailed discussion of environmental enforcement and the serious consequences associated with violating our nation's environmental laws. Finally, we have included several appendixes in the book that provide the environmental manager with practical reference material.

Environmental management is a dynamic mix of law, science, public policy, and communications. In this book, we supply the environmental manager with the basic tools for understanding and resolving the most common environmental management issues affecting the operation of industrial facilities in the United States. We hope the insights we gained from our experience with environmental management and legal issues will help environmental managers throughout the country meet the challenges of the next century.

ACKNOWLEDGMENTS

Our thanks go to several people who helped us complete this project. Ed Clem, Vice President of Environmental Affairs at Champion International, honored us by agreeing to review our work and prepare the comments noted in the Foreword. Dennis Ross of Boise Cascade Corporation reviewed the entire manuscript and gave us many helpful comments and criticisms. Roy Elliott and Ann Campbell at Dames & Moore assisted in developing several of the illustrations used in the book. In addition, Roy Elliott provided comments on several chapters. Sharon Martin, Debbie Stone, Karen Nimis, and the rest of the staff at Bogle & Gates P.L.L.C. provided continuous word processing and organizational support and patiently worked with us through numerous revisions. We also commend EPA for the wealth of excellent information made available to the public through its various websites and publications. In addition, state and local agencies, including the Washington Department of Ecology and California's South Coast Air Quality Management District, provided us with useful reference forms and tables. Finally, no acknowledgment would be complete without recognizing our wives, Jan Erickson and Nancy King, for their continuing support and encouragement.

1

ENVIRONMENTAL MANAGEMENT OVERVIEW

Managing the environmental program in an industrial facility has become an increasingly complex and challenging assignment. New regulations, increasing public expectations, and better scientific and analytical techniques are just a few of the areas that have acted synergistically to create the demanding world in which environmental managers must function. Today's environmental manager must have the scientific skills to monitor and test for compliance, the creative ability to diagnose complex problems related to the production process, a strong fundamental understanding of the applicable legal requirements, and the interpersonal skills to deal with facility management, regulatory agencies, and the general public. It is a wonder anyone can qualify for, let alone aspire to, the job. Yet some of the most dedicated and technically qualified employees in industry occupy these positions. For many, the satisfaction of maintaining a strong environmental program at an industrial facility apparently outweighs the rigors of the job. The intent of this book is to help those environmental managers do their work better by emphasizing the "how to" aspects of environmental management.

In most industrial facilities, the environmental management program has become a team effort by necessity. There are simply too many responsibilities for any one individual to handle. The job also requires full-time availability to address spills, upsets, and other emergency events. These demands have forced operations managers to become part of the environmental management team. This has had a positive effect on environmental compliance, as more managers within the organization become aware of the environmental regulatory requirements and the adverse consequences of noncompliance.

The foundation of environmental management is an understanding of the laws and regulations that apply to a facility. Industrial facilities are typically subject to a multitude of laws and regulations governing air, water, solid and hazardous waste, and chemical handling. These regulatory requirements can vary, depending on state requirements and whether the Environmental Protection Agency (EPA) has delegated authority to the state to administer the environmental regulatory program.

It also is important to remember that environmental liabilities go beyond the property boundaries of an operating facility. Environmental managers must ensure that any hazardous wastes shipped off-site are sent to reputable hazardous waste management facilities that will properly handle the waste. Managers must also inform the public of the potentially harmful releases of chemicals from their facilities, as required by the Community Right-to-Know Law and similar regulatory programs.

Knowing the laws and regulations that apply to the facility is only the first step in developing a successful environmental management program. Developing mutual respect and maintaining relationships with the environmental agencies and individuals who enforce the laws is also extremely important. Moreover, because most of the environmental laws contain citizen suit provisions that authorize the public to enforce the laws in the absence of agency action, environmental managers must inform and involve the public in a broad range of environmental issues.

Once a fundamental understanding of the laws, regulations, regulatory agencies, and interested citizens has been achieved, the environmental manager should develop site-specific programs for each area of environmental regulation. These areas tend to fall into the following four management or "media" groupings:

- Air
- Water
- Waste
- Chemicals

A successful environmental management system must include a program to ensure that the facility's ongoing operations comply with the law and that any changes in operations receive the necessary regulatory approvals or permits. Air quality permitting offers probably the best example of how changes in the operation of a facility can affect environmental permitting requirements. Proposed facility changes that result in air emission increases are almost always subject to some level of agency review and approval prior to implementing a process change or installing new equipment. The length of time required to obtain air permits has grown with the complexity of the environmental regulations, and permitting delays can be devastating to the success of a proposed process change or expansion project.

Once compliance programs have been established at a facility, they should be supported by good record keeping, a central filing system, and a key date compliance calendar. Agency inspectors and environmental groups often judge a facility by the completeness and accuracy of its environmental record keeping. The message is clear: Do not neglect the proper documentation of environmental compliance.

Environmental managers also have to become members of the overall business management team for the facility and participate in capital planning and approval decisions. Environmental management considerations must become a part of the day-to-day operations of the plant, and all facility managers and employees should understand the importance of environmental management to the success of the business.

How are you doing? Many companies supplement their overall facility programs by conducting oversight environmental audits. Individual facilities also conduct periodic environmental self-assessments. Both of these tools are used to improve the environmental management program at the facility and ensure that issues do not evolve into major compliance problems.

Should your facility go "beyond compliance"? The advent of programs such as ISO (International Standards Organization) 14000 have transformed this question from a local to a global issue. In the future, the environmental performance of industrial facilities around the world will be compared with the use of the same set of standards, and the ability of companies to meet these standards may affect the acceptability of their products in the marketplace. Many customers today ask for an "environmental certification" prior to purchasing a product. This suggests that programs such as Pollution Prevention and Life Cycle Assessment (LCA) will become integral components of future environmental management systems. This convergence of products, markets, and environmental management underscores the conclusion that environmental management at an industrial facility in the future will become the responsibility of all facility employees and a more important part of the overall management of the business.

We hope this book will help you meet the environmental management challenges of the next century.

2

INTRODUCTION TO ENVIRONMENTAL LAW

It is vitally important in environmental management to identify the sources of environmental law and define the respective roles of the state and federal government in administering the law. The environmental manager should develop a working knowledge of all the regulatory provisions applicable to the facility. Most federal environmental statutes give the Environmental Protection Agency (EPA) primary responsibility for enforcement but allow EPA to delegate its responsibilities to the states. The system is further confused by the fact that the enacting legislation often fails to create clear boundaries between the respective authorities of the federal government and the states. In general, a state may administer its own environmental program so long as the program requirements are as stringent as federal requirements. Many states have enacted environmental laws that are more stringent than the corresponding federal program, or environmental programs that are completely independent of any authorizing federal law. Chapter 28 of this book discusses in more detail the relationship between federal and state agencies.

This chapter focuses on the development of federal environmental law; however, state laws, regulations, and legal principles have similar origins.

2.1 STATUTES, REGULATIONS, AND EXECUTIVE ORDERS

Statutes passed by Congress and approved by the president are the primary sources of federal environmental law. Statutes vary in length from several pages for the National Environmental Policy Act (NEPA) to several hundred pages for the Clean Air Act (CAA). Federal statutes are published in the

United States Code (USC), which can be found in all law libraries and many public libraries. Several publishing companies also sell compilations of the federal environmental statutes, which are widely available and reasonably priced.

In general, environmental statutes identify the objectives of the law and give EPA broad authority to implement necessary regulatory programs. However, recent environmental statutes such as the Clean Air Act Amendments (CAAA) of 1990 have relieved the agency of much of its discretion. Instead, Congress, through hundreds of pages of legislation, establishes the specific details of the regulatory program. Federal environmental statutes are supported by legislative history, which is used by attorneys and courts to decipher the intent of Congress in passing a law.

Federal environmental regulations are initially proposed by EPA and published in the *Federal Register*. The public is given an opportunity to comment on a proposed regulation, and the agency is required to consider comments prior to finalizing the rule. Once regulations are finalized, they are published again in the *Federal Register* and subsequently codified in the Code of Federal Regulations (CFR). The regulatory preambles that are published with both the proposed and final rules in the *Federal Register* are an important source of information regarding the agency's intent in developing the rules. For the reader's convenience, the applicable CFR citations have been provided in this book at the beginning of each chapter dealing with specific federal regulations. A partial list of CFR sections relating to the environment is provided in Table 2.1.

The president of the United States occasionally issues executive orders relating to environmental matters. These orders, another source of federal environmental law, clarify procedures or establish policies for EPA and executive agencies to follow in addressing environmental matters. For example, Executive Order 12843 establishes federal procurement requirements and policies for ozone-depleting substances and Executive Order 12969 governs federal acquisition procedures as they relate to the Community Right-to-Know Law.

Environmental statutes, regulations, and executive orders are interpreted by courts, by administrative tribunals, and by the agencies themselves. Court decisions and administrative tribunal opinions are typically published and can be found in law libraries, through computerized legal research services, or on the Internet.

In addition to statutes, regulations, court decisions, and other formal legal materials, EPA issues a broad array of policy statements and guidance documents that an agency uses to administer the law. These materials typically are not readily available to the public and are not subject to formal public review and comment process, although agencies sometimes solicit comments

Table 2.1 Environmental regulations

Code of Federal Regulations (CFR) Citation	Scope of Coverage
33 CFR 151, 153–158	Oil spills
40 CFR 50–58, 60–63, 66–78, 81, 82, 87, 93, 95	Clean air (primary, secondary, new source, and hazardous emissions)
40 CFT 79, 80, 85, 86, 88–91, 600, 610	Motor vehicle emissions
40 CFR 21, 108–110, 112, 113, 116–117, 121–123	Water (general, oil, hazardous substances)
40 CFR 125, 130–133, 135, 136, 140, 501, 503	Water (NPDES, pretreatment standards)
40 CFR 141–149	Drinking water and underground injection
40 CFR 190–192, 194, 195	EPA radiation standards
40 CFR 201–205, 209–211	Noise abatement
33 CFR 320–330	Dredge and fill permits
40 CFR 238, 240, 243, 244, 246, 247, 254–258, 260–266, 268, 270–273, 279–282	Solid and hazardous waste
40 CFR 300, 302–305, 307, 310, 311	Superfund
40 CFR 350, 355, 370, 372, 374	Community Right-to-Know
40 CFR 104, 129, 401, 403, 405–415, 417–436, 439, 440, 443, 446, 447, 454, 455, 457–461, 463–469, 471	Effluent standards
40 CFR 1500–1508, 1515–1517	CEQ, NEPA, and selected general rules

from the regulated community. EPA's extensive use of informal policy statements when making administrative decisions has been widely criticized by the regulated community.

2.2 COMMON-LAW THEORIES OF LIABILITY

Although displaced in large degree by environmental statutes and regulations, there still exists a body of ''common law'' that can be used to address environmental grievances. The common law is a series of court-developed decisions that govern personal and property rights. The area of the common law that relates to the general legal duty to avoid causing harm to others is called ''the law of torts.'' The three types of torts typically used to address environmental issues are nuisance, trespass, and negligence. These tort theories are most often used by neighbors of industrial facilities who are offended by the odor, dirt, smoke, water pollution, or hazardous substance contamination allegedly caused by the neighboring facility. This area of the law is significant in that it is administered by the courts rather than the regulatory agencies, and juries are usually called upon to determine the monetary damages that should be paid to the plaintiffs.

2.3 CARDINAL PRINCIPLES OF ADMINISTRATIVE LAW

Although it is beyond the scope of this book to summarize all the ways in which the legal process affects the administration of environmental law, it is important for environmental managers to understand a few of the basic rules of administrative law. The procedural law established by the federal Administrative Procedures Act governs not only an agency's operations and promulgation of rules but, most important, defines when a court is authorized to overrule an agency decision. Some of the cardinal principles of administrative law are as follows:

- Agencies must act in accordance with either procedures established by the statute creating the law in question or the Administrative Procedures Act. Failure to follow these procedures typically will allow courts to invalidate agency action.
- An agency must maintain a record of its action, and there must be enough evidence in the record to support the agency's action. If an agency is taken to court, the court will focus on the "administrative record." The court will not decide whether the agency decision was right or wrong but, rather, whether the agency followed the procedures of the Act to reach a decision that is within its authority.
- Agency decisions can be reviewed by courts, but only if you have standing to sue and the agency has issued a "final" decision. Nor may you sue until you have "exhausted your administrative remedies"; that is, you must pursue your appeal through the established administrative process before going to court.
- Courts generally defer to an agency's expertise in interpreting its own statutes and regulations.
- Agency decisions are overturned by the courts only if such decisions are "arbitrary and capricious," "an abuse of discretion," or "otherwise not in accordance with law." In general, if there is some evidence in the record supporting an agency's decision, the court will not overrule it.

2.4 HISTORY OF ENVIRONMENTAL LAW

Although some laws addressing conservation matters and specific environmental issues have existed since the turn of the century, comprehensive federal regulation of industrial pollution was virtually unknown prior to the enactment of the National Environmental Policy Act of 1969 (NEPA). With the passage of NEPA, Congress established a broad environmental policy for the federal government and started a landslide of environmental legislation and regulation. NEPA was followed by the Clean Air Act of 1970, the Federal Water Pollution Control Act Amendments of 1972, the Endangered Species Act of 1973, the Resource Conservation and Recovery Act of 1976, the Toxic Substances Control Act of 1976, and the Clean Water Act of 1977. These statutes were accompanied by thousands of pages of implementing regulations and agency guidance.

The wave of environmental legislation continued throughout the next decade with the passage of the Superfund Law in 1980 and the Hazardous and Solid Waste Amendments of 1984. The Superfund Law was significantly amended in 1986 through the enactment of the Superfund Amendments and Reauthorization Act of 1986 (SARA). Title III of SARA, the Emergency Planning and Community Right-to-Know Act of 1986, required certain industrial facilities to establish effective emergency planning programs to protect the public from accidental releases of hazardous substances.

The major environmental law of the 1990s has been the Clean Air Act Amendments of 1990, which created a statutory and regulatory program that goes far beyond any of its predecessors in its scope and level of detail. In many respects, the federal EPA and its state counterparts are still struggling to meet the deadlines and responsibilities established by the environmental statutes passed by Congress during the last 20 years. Federal and state agen-

Table 2.2 Chronology of major modern laws governing environmental matters

Initial Passage	Act
1990	The Oil Pollution Prevention Act of 1990
	The Pollution Prevention Act
1986	The Superfund Amendments and Reauthorization Act
	The Emergency Planning and Community Right-to-Know Act
1980	Superfund
1977	The Clean Water Act
1976	The Resource Conservation and Recovery Act
	The Toxic Substances Control Act
1974	The Safe Drinking Water Act
1973	The Endangered Species Act
1972	The Federal Insecticide, Fungicide, and Rodenticide Act
1970	The Clean Air Act
1969	The National Environmental Policy Act

cies are chronically too understaffed and underfunded to fulfill many of the environmental mandates specified by Congress. Nevertheless, the introduction of new environmental legislation is inevitable, and environmental managers must continue their efforts to influence new legislation, interpret it, and apply it to their facilities. Table 2.2. shows all of the landmark federal environmental legislation passed in the last 30 years.

3

SUMMARY OF THE CLEAN AIR ACT

Statute:	42 USC§§ 7401 to 7671
Regulations:	40 CFR Parts 50 to 95

The Clean Air Act (CAA) is a comprehensive federal statute designed to regulate air emissions from stationary and mobile sources. Major federal clean air legislation includes the Air Quality Act of 1967, the Clean Air Act Amendments of 1970, the Clean Air Act Amendments of 1977, and, most recently, the Clean Air Act Amendments of 1990.

The 1990 Amendments dramatically changed the regulation of air pollution and required EPA to promulgate hundreds of new regulations. Consequently, the CAA is now evolving faster than any area of environmental law. Some of the key concepts necessary to understand the operation of the CAA are summarized in this chapter. Several of these subjects also are discussed in further detail in later chapters of the book.

3.1 AMBIENT AIR QUALITY STANDARDS

One of the first steps toward improving air quality in the United States was the Environmental Protection Agency's (EPA) establishment of ambient air quality standards. Ambient air quality standards are national in scope and set air quality goals to be achieved by the states. The standards are stated in terms of annual concentration levels or annual mean measurements for the air.

Based on the criteria established for each pollutant under Section 108(a) of the Clean Air Act and "allowing an adequate margin of safety," EPA is directed to promulgate national primary and secondary ambient air quality standards for each pollutant. Primary air quality standards are designed to protect the public health. Secondary standards are intended to protect the public welfare from any known or anticipated adverse effects associated with the presence of such pollutants in the air. Primary and secondary standards have been established by EPA for several pollutants. The following are the major criteria pollutants:

- Carbon monoxide
- Nitrogen oxides
- Sulfur dioxide
- Particulate matter
- Ozone
- Lead

3.2 STATE IMPLEMENTATION PLANS (SIPs)

National air quality regulations are applied to individual sources through State Implementation Plans (SIPs). A SIP is an extensive, detailed document that contains elements such as emission inventories, monitoring programs, attainment plans, and enforcement programs. Section 110 of the CAA requires a SIP to include all the information necessary to implement the Act within the state. The SIP also imposes whatever specific emission limitations are required to meet national ambient air quality standards within each air quality region of the state. The SIP is a living document, which is revised by the state environmental agency on an ongoing basis. If a state fails to adopt an acceptable SIP, EPA is directed to formulate and enforce one for that state.

3.3 NEW SOURCE PERFORMANCE STANDARDS (NSPS)

Performance standards for new sources are designed to allow industrial growth without undermining the national program for achieving air quality goals. New Source Performance Standards (NSPS) are established at the national level in order to prevent states from becoming "pollution havens" and attracting industry with their lenient emission standards.

NSPS differ from ambient standards in both purpose and form: NSPS are oriented to particular sources of pollutants rather than to air quality in general. NSPS are typically numeric standards that relate to the level of pollution control achieved by installing the best demonstrated technology. NSPS are examined in further detail in Chapter 8.

3.4 PREVENTION OF SIGNIFICANT DETERIORATION (PSD)

The Prevention of Significant Deterioration (PSD) program has grown out of requirements of the Clean Air Act. The program is designed to prevent significant deterioration of air quality in regions where the air is already cleaner than mandated by ambient standards.

To implement the regulatory program for PSD, all air quality control regions are designated by class for the purpose of specifying the amount or "increment" of air pollution that can be permitted in each area.

Generally speaking, the PSD regulations apply to the construction or modification of a major source of any pollutant located within an area designated "attainment" or "unclassifiable" for any criteria pollutant (see Section 6.1). To obtain a PSD permit, the source must

- Submit required monitoring data
- Apply best available control technology (BACT) to each pollutant that the source emits in more than the minimum amounts
- Model the projected emissions to demonstrate the increment of ambient air quality consumed by the proposal and demonstrate future compliance with ambient air quality standards
- Undergo agency review and public hearing

3.5 NONATTAINMENT AREAS

Nonattainment areas are air quality control regions that have not met the ambient air quality standards. EPA allows new major sources in nonattainment areas only if stringent conditions are met, including a greater than one-for-one offset of emissions from existing sources in the area.

The emissions offset policy was adopted as part of the 1977 Amendments to the Clean Air Act to allow industrial growth in areas currently violating the ambient air quality standards and to ensure continued progress toward attainment of the standards. EPA's present policy also allows development of a procedure to permit the "banking" (saving) of offsets. Excess emission reductions, beyond the point needed for immediate trade, can be banked for use in future offset trades. Banked emission reductions become assets of the owner company.

The offset policy means that a proposed source wishing to locate in a nonattainment area must negotiate enforceable agreements from existing sources for emission reductions greater than the amount of its expected emissions. The proposed source must also meet emission limitations based on the Lowest Achievable Emission Rate (LAER) for such a source and must certify that all other sources owned by it are in compliance with SIP requirements.

3.6 HAZARDOUS AIR POLLUTANTS (HAPs)

Section 112 of the Clean Air Act is EPA's mandate to control hazardous pollutants discharged into the nation's air. An agency is authorized under this section to promulgate National Emission Standards for Hazardous Air Pollutants (NESHAPs) for both new and existing sources. EPA's past activities in promulgating NESHAPs for hazardous air pollutants such as asbestos were widely perceived as a major failure. Consequently, the 1990 Amendments established an initial list of 189 hazardous air pollutants (HAPs) and gave EPA the authority to periodically review and add HAPs to the list.

EPA is also required under the amendments to promulgate technology-based limitations for industrial source categories and issue standards for each category. It is estimated that as many as 250 categories will be established, with major sources subject to maximum achievable control technology (MACT) limitations that are determined by EPA.

EPA is further required to develop a strategy to reduce human cancer risks from area source emissions by at least 75%. Either Congress or EPA must impose additional controls, as needed, to protect the public health or prevent significant, widespread environmental harm and provide an ''ample margin of safety.'' Moreover, if the lifetime residual risk of cancer from the remaining HAPs is not reduced to less than one in one million, provisions to control known carcinogens will be automatically triggered.

In addition to a new HAPs program, a comprehensive program for accidental hazardous substance release prevention, reporting, and investigation has been established. The owners/operators of stationary sources have a ''general duty'' to identify the hazards of accidental releases, take steps necessary to prevent releases, and minimize the consequences of accidental releases.

3.7 VISIBILITY PROTECTION FOR FEDERAL CLASS I AREAS

Section 169A of the Clean Air Act requires visibility protection for mandatory Class I federal areas where it has been determined that visibility is an important value. Mandatory Class I federal areas are all international parks and certain national parks and wilderness areas that are identified in Section 162 of the Clean Air Act.

3.8 AIR OPERATING PERMITS

Title V of the 1990 Clean Air Act Amendments establishes a detailed system for issuing air operating permits to most major sources of air pollution. In general, these permits will include emissions limitations, schedules of compliance, monitoring requirements, and compliance certifications. Most air operating permit programs will be administered by the authorized state.

3.9 MOBILE SOURCES

The Clean Air Act has regulated air emissions from mobile sources (primarily motor vehicles) for decades. These provisions of the Act apply almost exclusively to vehicle makers and fuel manufacturers. The 1990 Amendments to the Act required vehicle manufacturers to further reduce emissions, and fuel refiners to develop new and reformulated fuels to achieve emission reductions.

3.10 ACID RAIN CONTROL PROGRAM

The 1990 Clean Air Act Amendments established another new program that requires a nationwide reduction of sulfur dioxide (SO_2) and nitrogen oxide (NO_x) emissions from fossil fuel–fired combustion devices that produce electricity for sale. A complex system of ''allowances,'' which, in essence, can be bought and sold, will be used to regulate SO_2 emissions from the electric industry.

3.11 STRATOSPHERIC OZONE PROTECTION

Title VI of the Clean Air Act creates a framework to regulate and eventually phase out the production of all chemicals that deplete the ozone layer. Title VI implements the treaty obligations imposed on the United States by the 1987 Montreal Protocol on Substances That Deplete the Ozone Layer. A detailed discussion of this program is found in Chapter 9 of this book.

4

AIR EMISSION INVENTORY AND ANALYSIS

The first requirement in determining the applicable air regulations and permit requirements for an industrial facility is the creation of an inventory of air pollutants emitted from the facility. For facilities that are major sources of emissions (generally defined as sources with more than 100 tons/year of criteria pollutants, or 10 tons/year of an individual hazardous air pollutant (HAP), or 25 tons/year of total HAPs), the inventory is also used to determine fee requirements for air operating permits. In the case of permitting, the inventory is used to establish an air emissions baseline for evaluating future changes.

4.1 IDENTIFYING SOURCES

To begin the inventory process, you should list all potential sources of air emissions regardless of the level of anticipated emissions. Some sources may be eliminated as *de minimis* contributors or unregulated sources, based on state or local regulations, at a later point in the analysis. Air pollutants are emitted either from stationary sources through stacks, chimneys, or vents, or as fugitive air emissions. Fugitive emissions are typically defined as any air emissions that cannot reasonably pass through a stack, chimney, vent, or other functionally equivalent opening. Examples of fugitive emissions include the emissions associated with the loading and unloading of ore in a mine pit and the volatile emissions released from the surface of an effluent treatment system. Regulatory and air modeling requirements for source emissions and fugitive emissions differ, requiring that the inventory list be divided by type of source.

Most existing industrial facilities will have completed an update of the source inventory in accordance with requirements under the 1990 Clean Air Act Amendments for air operating permit applications. However, new facilities and proposed modifications to existing facilities also require the preparation of a source inventory. In these cases, because the source has not yet been constructed, there is no existing emission information to review. The emission inventory and points of emission will be based on information obtained from resources such as industry associations, process flow diagrams, scale models, and/or visits to similar facilities.

When you are preparing an inventory for a new or modified facility, it is important to maintain good communications with the engineering design team for that project. The environmental manager will often serve as the point of contact between the permit team and the project team. It is highly likely that the number and type of emission sources will change from the initial inventory list, which is typically completed early in the design stages of the project. Final emission estimates will have to be included in permit

Table 4.1 Building heights of structures

Structure No.	Height (ft)	Height (m)
1	40.0	12.2
2	10.0	3.1
3	8.0	2.4
4	8.0	2.4
5	40.0	12.2
6	10.0	3.1
7	12.0	3.7
8	34.1	10.4
9	25.0	7.6
10	25.0	7.6
11	8.0	2.4
12	50.0	15.2
13	32.0	9.8
14	62.0	18.9
15	49.9	15.2
16	54.1	16.5
17	30.0	9.1
18	30.0	9.1
19	35.1	10.7
20	46.0	14.0
21	40.0	12.2
22	60.0	18.3

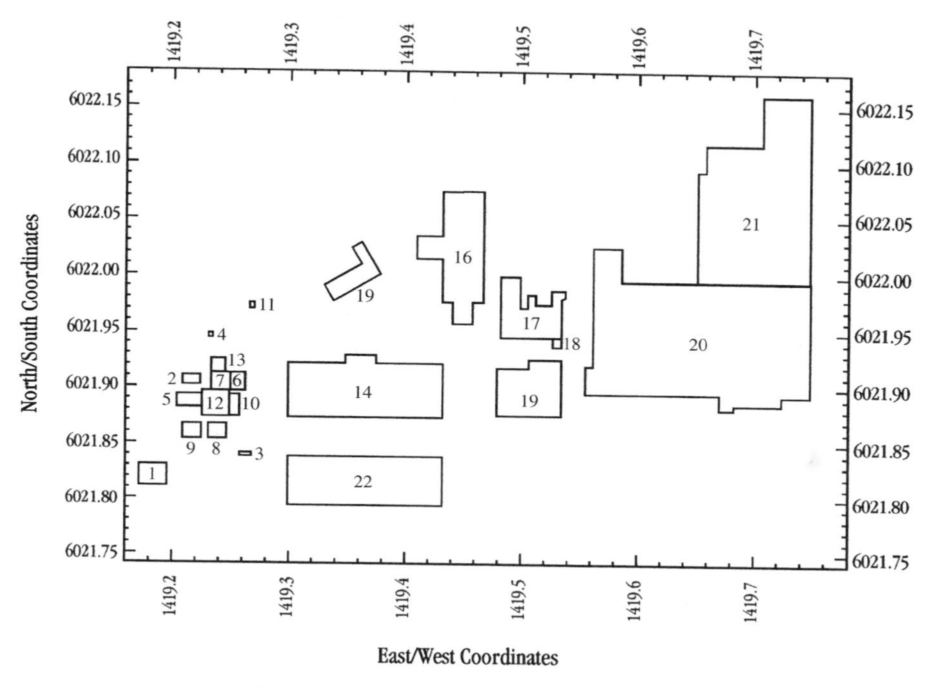

Figure 4.1 Project structure locations

applications before agency approval can be obtained. Factual misunderstandings between the project team and the permit team are among the primary causes of delay in obtaining the necessary regulatory approvals.

If air modeling is part of the permitting process, the coordinates and elevation of the emission sources and the heights and coordinates of surrounding buildings should be added to the source list. Table 4.1 is an example of a chart identifying building heights for various structures. Figure 4.1 identifies the locations of these structures. Air modeling is required to demonstrate compliance with ambient air quality standards or determine levels of hazardous air pollutants at the property line.

4.2 ACTUAL VERSUS POTENTIAL EMISSIONS

Once a list of sources is established, the facility can begin estimating its air emissions. The following two emission estimates are typically needed:

- *Potential Emissions*. Emissions associated with the maximum capacity of a stationary source to emit any air pollutant under its physical and operational design. Any physical or operational limitations on the capacity of the source are considered in establishing potential emissions if the limitations are enforceable by the authorized regulatory agency.

- *Actual Emissions*. Measured mass emissions of a pollutant from an emissions source during a specified time period.

Several scenarios demonstrate the difference between potential and actual emissions. One occurs when the normal production or use of the emission source is less than the designed nameplate capacity of the source. Another happens when market demands reduce the hours of source operation (e.g., five-day operation rather than a normal seven-day schedule).

In permitting a facility, try to avoid permit provisions that limit the hours of plant operation. If a production rate has been included as an enforceable permit limitation, the limit cannot be exceeded until a permit modification is approved by the regulatory agency.

4.3 ESTIMATING EMISSIONS

The list of federally regulated criteria pollutants that must be estimated in the emissions inventory are summarized in Table 4.2; Table 4.3 identifies common HAPs. An emission inventory should include known HAP emis-

Table 4.2 New Source Review (NSR) pollutants

Pollutant	Regulated Under*
Sulfur dioxide	NAAQS, NSPS
Particulate matter (TSP)	NSPS
Particulate matter (PM_{10})	NAAQS
Nitrogen oxides	NAAQS, NSPS
Carbon monoxide	NAAQS, NSPS
Volatile organic compounds (ozone)	NAAQS, NSPS
Lead	NAAQS
Sulfuric acid mist	NSPS
Total fluorides	NSPS
Total reduced sulfur	NSPS
Reduced sulfur compounds	NSPS
Hydrogen sulfide	NSPS
Asbestos	NESHAP
Beryllium	NESHAP
Mercury	NESHAP
Vinyl chloride	NESHAP

*NAAQS = National Ambient Air Quality Standards, NSPS = New Source Performance Standards, NESHAP = National Emission Standards for Hazardous Air Pollutants

Table 4.3 Common hazardous air pollutants

Asbestos	Mercury compounds
Cadmium compounds	Methanol
Chlorine	Nickel compounds
Chloroform	Pentachlorophenol
Chromium compounds	Phosphorous
Cyanide compounds	Radionuclides
Formaldehyde	Selenium compounds
Hydrochloric acid	Toluene
Lead compounds	Xylenes
Manganese compounds	

sions as well as criteria pollutants. It is important to note that state and local laws may regulate pollutants that are not on the federal lists.

The list of regulated pollutants should be analyzed to see which pollutants are emitted from each source. This is easier said than done in most cases, especially in regard to the list of HAPs. For example, emissions once considered as a group to be volatile organic compound (VOC) emissions may, in fact, contain specific HAPs. Several methods are available to determine which pollutants at what quantities must be evaluated to ensure regulatory compliance.

Manufacturer data is available for many emission sources or production processes. Equipment built by different manufacturers often emit different levels of pollutants. A conversation with equipment suppliers can provide verification of the pollutants emitted and the estimated quantities of emissions. These estimates may be based on such criteria as production rate, fuel input, chemical feed rate, or bulk handling rates. Always verify the intended operating parameters with the operations manager or, in the case of new equipment, the design team for the particular production unit.

The EPA has summarized emissions data for many production units in various industries in a document entitled ''AP-42.'' The AP-42 document can be a very good source of emissions data if a specific equipment manufacturer has not yet been selected. However since the AP-42 document contains a broad range of data for some sources, selecting a representative emission rate for your equipment may be challenging. AP-42 emission rates should only be used if there is a strong correlation with the actual equipment under consideration. Internal information from your company or organization is also a very common source of emissions data. Similar equipment may be used at other company facilities. Before relying on information on emissions from similar equipment, compare the specifications of the new source with the production rates and process chemistry of the existing equipment. For example, a small change in fuel oil sulfur content can lead to a major change in SO_2 emissions from a similar steam boiler.

If all else fails, source testing may be required. In the case of HAPs, source testing may be the only reliable source of information owing to the overall lack of HAP data. However, before you decide to test, ensure that an acceptable test protocol exists. Not all of the chemicals listed as HAPs have scientifically agreed-upon test protocols. Even VOC testing can be difficult. Some VOCs are condensable and become aerosol particulates at cooler temperatures. New regulatory standards for emitted particulate matter that has a diameter no greater than 2.5 microns ($PM_{2.5}$) will also require test protocol development. Until that protocol is developed, it appears that PM_{10}, particulates no greater than 10 microns in diameter, will continue to serve as the particulate measure to be included in the inventory.

Source test data must be reported to the authorized agency. Because test results are often transformed into permit limitations, you should determine prior to testing whether the test will produce acceptable data. Emission tests should be conducted by qualified testers and carefully designed to demonstrate rated capacity conditions. The source should be operated properly during the test. Do not test a source you know is malfunctioning or operating well below design conditions, because the tests results will not reflect the typical operation of the source.

4.4 CREATING THE INVENTORY

An emissions inventory is created with the use of a list of sources, emission data, and production data. Table 4.4 presents a typical emissions inventory summary, showing the increases and decreases in criteria pollutant emissions

Table 4.4 Summary of net controlled emission change due to a proposed project

Equipment	CO (tons/year)	NO_x (tons/year)	SO_2 (tons/year)	VOC (tons/year)	PM (tons/year)
Gas Boiler	21.40	20.24	0.58	9.25	5.78
Flash Dryer				75.20	8.04
Boiler (post-modification)	119.45	130.90	6.50	19.08	100.60
Boiler (existing)	(802.44)	(88.50)	(4.43)	(12.98)	(68.44)
Eliminated Sources	(1292.17)	(22.42)	(4.43)	(10.62)	(203.56)
Product Dryers (existing)				(31.75)	(2.52)
Product Dryers (post-modification)				52.41	4.17
Net Emission Change	(893.77)	39.40	(1.77)	100.60	(155.94)

relating to a proposed modification project. In this example, the environmental manager will have to evaluate the regulatory significance of the projected increases in NO_x and VOC emissions. The increase in VOC emissions will likely require the source to undergo new source review before an air emissions permit authorizing the project can be issued.

4.5 DETERMINING PERMIT REQUIREMENTS

Once you have completed the air emissions inventory, an analysis of the results will determine which requirements of the Clean Air Act are triggered. Three different types of air permits can be required:

- *Title V Air Operating Permit (AOP).* If you are analyzing an existing or proposed facility that exceeds 100 tons/year of a criteria pollutant or emits a single hazardous air pollutant in excess of 10 tons/year (or 25 tons/year of combined HAPs), you will likely have to obtain a Title V permit. A source subject to New Source Review also is likely to need an AOP.
- *Permit to Construct (PTC) or Notice of Construction (NOC).* If you are proposing a new facility, or the modification of an existing facility, that will result in a moderate increase in air emissions, your state and local rules likely will require you to obtain an air permit prior to construction. These requirements vary from state to state and should be determined for your particular location. Small sources may be exempt from this requirement or subject to a streamlined permitting procedure.
- *New Source Review (NSR) or Prevention of Significant Deterioration (PSD) Permits.* The construction or modification of a major stationary source that results in "significant" increases in air emissions is subject to the New Source Review program established by the Clean Air Act.

All of these categories of permits and the related application procedures are described in more detail in the following chapters. The environmental manager should note that requirements for Air Operating Permits are independent of those for construction permits. You may determine that both types of permits are required for your project.

5

PERMITS TO CONSTRUCT

A facility planning to construct a new air emission source or modify an existing source will most likely be subject to some level of air permitting. New or modified emission sources whose increased emissions are below established thresholds are typically required to apply for what is commonly known as a Permit to Construct (PTC). Some states use other terms such as Notice of Intent to Construct (NOI) or Notice of Construction (NOC). As with all New Source Review permits, the permit must be approved by the authorized agency prior to commencing construction of the source. This approval process may take several months and should be factored into any construction planning and operation schedules. PTC requirements vary from site to site inasmuch as they are governed by state and local rules.

Some states provide for *de minimis* exemptions, which may preclude the need for an application or require only a notification letter informing the agency of the planned activity. For example, small boilers (<10,000 Btu/hr heat input) for building heat are often subject to a registration process rather than a full permit review. Pilot and research projects are also often exempted. As a result of the Title V Air Operating Permit program, many states have reviewed and modified their construction permitting programs to correspond with Title V requirements.

The examples used in this chapter are based on the application requirements of the South Coast Air Quality Management District in Southern California. The actual requirements in your area should be reviewed with your local environmental agency.

5.1 PTC GENERAL REQUIREMENTS

General requirements for a Permit to Construct are ordinarily specified in an agency form. An example form from the South Coast Air Quality Manage-

South Coast Air Quality Management District
21865 East Copley Drive
Diamond Bar, CA 91765
(909) 396-2000

APPLICATION FOR PERMIT TO CONSTRUCT AND PERMIT TO OPERATE
FORM 400 - A

NC/NOC NUMBER:
INSPECTOR SECTOR
ISSUE DATE

Company Information

LEGAL NAME OF APPLICANT ☐ IRS OR ☐ S.S.NUMBER
PERMIT TO BE ISSUED TO (SEE INSTRUCTIONS)
BUSINESS MAILING ADDRESS
TYPE OF ORGANIZATION
☐ Corporation ☐ Limited Partnership ☐ Government Entity
☐ Individual ☐ General Partnership ☐ Other:
ARE YOU A SMALL BUSINESS? (SEE INSTRUCTIONS) ☐ Yes ☐ No | AVERAGE ANNUAL GROSS RECEIPTS | NUMBER OF EMPLOYEES | IS YOUR BUSINESS 51% OR MORE WOMAN/MINORITY OWNED? ☐ Yes ☐ No
ARE ALL FACILITIES UNDER SAME OWNERSHIP IN CALIFORNIA IN COMPLIANCE WITH FEDERAL, STATE, AND LOCAL AIR POLLUTON CONTROL RULES? ☐ Yes ☐ No
ARE YOU THE OWNER OF THE EQUIPMENT UNDER THIS APPLICATION? ☐ Yes ☐ No ☐ IRS OR ☐ SS. NUMBER OF OWNER
IF NO, ENTER THE LEGAL NAME OF OWNER

Facility Information

EQUIPMENT ADDRESS/LOCATION | FACILITY NAME
NUMBER/STREET | FACILITY ID NUMBER
CITY OR COMMUNITY CA, ZIP CODE
NAME OF CONTACT PERSON AND TITLE
TYPE OF BUSINESS AT THIS FACILITY | NUMBER OF EMPLOYEES AT THIS FACILITY
CONTACT TELEPHONE ( ) - | BUSINESS TYPE CODE (SEE INSTRUCTIONS) | IS THERE A SCHOOL WITHIN 1,000 FEET OF YOUR PROPERTY? ☐ Yes ☐ No

Equipment Information

EQUIPMENT DESCRIPTION (SEE INSTRUCTIONS)
APPLICATION FOR (SEE INSTRUCTIONS)
☐ New Construction ☐ Modification
☐ Change of Location ☐ Change of Permittee
☐ Existing Equipment With Expired Permit ☐ Change of Permit Condition
☐ Existing Equipment Operating Without a Permit
ARE YOU SUBMITTING MULTIPLE APPLICATIONS FOR EQUIPMENT IDENTICAL TO THAT DESCRIBED ABOVE? ☐ Yes ☐ No
HAVE YOU BEEN ISSUED A NOTICE TO COMPLY (NC) OR A NOTICE OF VIOLATION (NOV) FOR THIS EQUPMENT? ☐ Yes ☐ No NC NUMBER: NOV NUMBER: ISSUE DATE: / /
HOW MANY PERMANENT OF FULL TIME JOBS WERE CREATED BY THE ADDITION OR MODIFICATIO OF THIS EQUIPMENT:
IF THIS EQUIPMENT HAS A PREVISOUS WRITTEN PERMIT, STATE NAME OF PERMITTEE | PREVIOUS PERMIT NUMBER
FOR NEW CONSTRUCTION OR MODIFICATION, ENTER ESTIMATED COST OF: BASIC EQUIPMENT $ AIR POLLUTION CONTROL EQUIPMENT: $
FOR NEW CONSTRUCTION OR MODIFICATION? ESTIMATED START DATE: / / ESTIMATED COMPLETION DATE / /
FOR CHANGE OF PERMITTEE, LOCATION, OR CONDITION ENTER DATE OF OCCURRENCE: / / | FOR EXISTING EQUIPMENT IN OPERATION WITHOUT PRIOR PERMIT, ENTER INITIAL OPERATION DATE / /
FOR THIS PROJECT HAS A CALIFORNIA ENVIRONMENTAL QUALITY ACT (CEQA) DOCUMENT BEEN REQUIRED BY ANOTHER GOVERNMENTAL AGENCY? ☐ Yes ☐ No
IF YES, ENTER NAME OF AGENCY AND SUBMIT A COPY IF APPROVED
DO YOU CLAIM CONFIDENTIALITY OF DATA? (SEE INSTRUCTIONS) ☐ Yes ☐ No
I HEREBY CERTIFY, UNDER PENALTY OF PERJURY, THAT ALL INFORMATION CONTAINED HEREIN AND INFORMATION SUBMITTED WITH THIS APPLICATION ARE TRUE AND CORRECT.
SIGNATURE TITLE OF SIGNER
TYPE OR PRINT NAME OF SIGNER | TELEPHONE NUMBER ( ) - | DATE: / /

SCAQMD USE ONLY	APPLICATON NUMBER	TYPE B C D	EQUIPMENT CATEGORY NUMBER	ASSIGNMENT UNIT ENGINEER	CLASS I II III	ENF. SECTION
	ENGR. INITIAL DATE / / A R	ENGR. INITIAL DATE / / A R	FEE SCHEDULE $	VALIDATION	CHECK OR MONEY ORDER NUMBER	AMOUNT $

43 Permits To Construct
\Chapter 05.doc 07/08/98

Figure 5.1 Application for Permit to Construct and Permit to Operate (Form 400-A) (Reprinted with permission of South Coast Air Quality Management District, Diamond Bar, California).

Table 5.1 Elements of a Permit to Construct application

- Company information
- Facility information
- Project description
- Control equipment description
- Cost of equipment
- Schedule information
- Application fee
- Signature of applicant

ment District has been included as Figure 5.1. PTC applications typically include the elements listed in Table 5.1.

5.2 SPECIFIC APPLICATION REQUIREMENTS

A PTC application requires the applicant to supply detailed information on the proposed new source, modification of an existing source, or process change. The required information can be obtained from the air emission inventory and analysis process described in Chapter 4. The common requirements of the PTC are summarized in the following sections.

5.2.1 Process Description

A narrative description of the proposed process change or modification must be prepared. The description should explain both the current operating scenario and the proposed change or addition and include production rates, hours of operation, types of emissions, and the sources of emissions. When discussing hours of operation, be sure to include the maximum production plan rather than the initial production or start-up procedures. The process description should also include all equipment changes and proposed emission control equipment that will be installed when the project is completed. A simplified process flow diagram can be a helpful tool in describing the project, as it offers a convenient way to display basic heat and material flows and quantities of emissions.

5.2.2 Site Plans

Site plans that show the facility layout and the location of new or modified equipment should also be included in the application. Site plans often require more than one level of detail, with one plan showing the overall facility and a second plan depicting the layout of the specific project. Site plans need not include detailed construction requirements. However, the better the agency understands the proposal, the faster it can process the application.

5.2.3 Equipment List

The equipment proposed for installation should be listed with a process description and, if available, manufacturer's data for rating production, heat input, and other relevant factors. The list should include design and normal production capacities of the equipment and any other information that may be used to prepare production or emission calculations. If the specific equipment has not yet been selected, use of a worst-case scenario will allow flexibility during the equipment selection process.

5.2.4 Air Emissions Inventory

A summary table of air emissions should be included in the PTC application to show increases (or decreases) in both criteria pollutants and hazardous air pollutants. Table 5.2 illustrates a simple inventory for the addition of a product labeler in a facility's shipping department.

5.2.5 PTC Appendixes

Normally, the goal of a permit applicant is to expedite approval of the permit. Clear and complete applications are much more likely to be approved by the agency in a timely manner. A thorough application typically includes appendixes that address the following topics:

- Emission calculations
- Equipment data sheets
- Process flow diagrams
- Screening concentration models for hazardous air pollutants

Table 5.2 Emission summary for proposed labeler addition (tons/year)

Source	PM	VOC	NO_x	SO_2	CO	HAP
Natural Gas	—	0.02	2.14	—	1.01	—
Ink	—	0.75	—	—	—	—
Glue	—	0.06	—	—	—	—
Total	—	0.83	2.14	—	1.01	—

Although not all of this information may be required by the applicable regulations, these materials will always be helpful to the reviewing agency and should be included if they are available.

5.3 OTHER POTENTIAL PTC APPLICATION REQUIREMENTS

Some state rules require that a PTC be accompanied by a simplified Environmental Impact Statement (EIS) or site review application. The terminology used to describe this requirement varies by state. For example, in the state of Washington, a State Environmental Policy Act (SEPA) checklist must be submitted with the PTC application. The SEPA checklist is designed to provide the agency and the public with a better idea of all the environmental impacts associated with the project. A copy of the Washington State SEPA checklist form is included in Appendix E.

Some states also require the applicant to submit an air dispersion model analysis to demonstrate that air quality will not be significantly impacted by the proposed project. In this case, more information about building heights and orientation may be required to complete the PTC. Finally, at least one state requires PTC applicants to complete a Best Available Control Technology (BACT) review for criteria pollutants for any project that increases air emissions more than over 1 ton/year. Although this requirement is not typical, it demonstrates the importance of reviewing the local rules before determining the final scope of the PTC application.

6

PERMITTING MAJOR SOURCES—PSD AND NSR

Regulation: 40 CFR Part 52, Subpart A

If a company or facility is planning to install a new air emission source that will result in significant emissions of a criteria pollutant, the proposed facility most likely will be classified as a new ''major source'' of air emissions. In addition, a proposed modification to an existing facility that will cause significant increases in the emissions of criteria pollutants will also be subject to major source review. Permitting of significant increases of air emissions from new or modified major sources is divided into two distinct categories. Prevention of Significant Deterioration (PSD) permitting applies to facilities that are located in air sheds that currently meet ambient air quality requirements (attainment areas). If a proposed project is located in an area that does not meet ambient air quality standards (nonattainment areas) or has consumed the increases allowed under the PSD program, the permit review process is typically called New Source Review (NSR). NSR requirements are usually the most stringent of all air regulations and are often difficult to meet.

Both PSD and NSR permitting involve a significant amount of documentation, air dispersion modeling, and regulatory agency review. The proposed permit will also be subject to public comment and review by federal land managers and, in the case of a significant or controversial project, one or more public hearings. The permitting agency, normally a state environmental agency, will consider and respond to public comments and possibly choose to modify the permit, based on those comments, prior to permit issuance. When permitting a major source, the facility permitting team should include in its schedule a significant amount of time for environmental agency review and response to comments. In the event the permit is revised as a result of these comments, the project team will have to determine whether the changes are acceptable or whether the permit changes cause the project to lose its financial viability owing to increased emission control costs.

Depending on the complexity of the project and the ambient air quality status in the vicinity of the proposed facility, agency review of a project proposal can take from six months to as long as several years. It is extremely important to understand the permitting requirements and expectations of the regulatory agency. Constant and continuing communications among the project design team, the environmental permitting team, and the environmental agency is the key to permitting success.

The quality of the initial PSD or NSR application can also affect the timely review of your application by the regulatory agency. Most environmental managers do not have all of the technical skills required to prepare a complete application. For that reason, environmental consultants are often hired to assist the permitting team, especially in regard to the computer models used to predict air emissions associated with the proposed project. When selecting an air permitting consultant, ascertain which individuals will be assigned to your project and verify that they have the required experience and a good working relationship with the lead regulatory agency.

Determining the lead agency will depend on what agency has primary responsibility for the air program in your state. In most cases, the federal Environmental Protection Agency (EPA) has delegated major source permit authority to the state environmental agency or a regional air authority. Regardless of which is the lead agency, EPA regional offices typically review and approve all major air permit applications. EPA also has authority to appeal PSD or NSR permits that, in its opinion, do not meet federal requirements. EPA intervention in permit issuance can lead to frustrating schedule delays if EPA concerns have not been addressed early in the permitting process. Theoretically, the EPA comments are directed to the state or local environmental agency, but the permittee is often required to respond directly to EPA's comments. This is especially true when financial determinations for proposed control technology are being questioned.

6.1 ATTAINMENT AND NONATTAINMENT AREAS

It is important to determine the ambient air quality status of the region in which a facility will be constructed to identify the set of requirements that apply to the proposed facility. The air quality designation of an area for each pollutant can be obtained from the local air authority and will be identified as either of the following:

- *Attainment.* Any area that meets the national primary or secondary ambient air quality standard for the pollutant.
- *Nonattainment.* Any area that does not meet the national primary or secondary ambient air quality standard for the pollutant.

If the air quality designation of an area for a particular pollutant is attainment, the proposed facility will be subject to the requirements of the Prevention of Significant Deterioration (PSD) program. Under the rules of PSD permitting, you must proceed with air modeling to determine whether you will increase the concentration of the pollutant beyond acceptable levels. Each pollutant in a given attainment area has a maximum allowable ambient concentration.

The project development team must be fully aware of the dynamic nature of the PSD process. First, you will have to determine what control technology or process design specifications are required to meet the ambient air quality standards. Second, a facility subject to the PSD program must install Best Available Control Technology (BACT). The determination of BACT is typically an iterative process, and changes in the technologies that meet the BACT definition commonly occur during the course of the PSD permitting process. Moreover, the permittee and the agency often disagree in their interpretations of BACT. Whenever BACT is changed, project design requirements, equipment selection, and the overall project schedule must also be revised.

When the air quality designation of an area for the pollutant you are increasing is classified as "nonattainment," the permit application will be subject to a much more stringent review under the NSR rules. Instead of BACT, the facility will be required to meet the "lowest achievable emission rates" (LAER). It also will be necessary to establish emission offsets for the proposed project that demonstrate that a net improvement in air quality will result from the construction of the project. For example, your project may require the installation of new pollution controls at an adjacent facility (possibly owned by someone else) to offset the proposed increase in the emissions of the nonattainment pollutant from your facility. The NSR process is fraught with uncertainty. Emission offsets are often unavailable, or the proposed project may not be economically feasible if high-cost LAER control technology is required. As a result of this uncertainty, projects proposed for nonattainment areas are frequently either abandoned or relocated to attainment areas once these issues are fully understood.

6.2 NEW OR MODIFIED SOURCE DETERMINATIONS

To determine whether a proposed project is subject to PSD or NSR review requirements, the initial step is to establish an emissions inventory. This process is discussed in detail in Chapter 4. Once an emission inventory is complete, data on the specific annual emissions of each pollutant will be available. A new facility with 100 tons/year of potential emissions of any air pollutant is classified as a major source if it is in the following federal list of stationary sources (see state air rules for additional sources):

- Fossil fuel–fired steam electric plants (>250 million Btu/hr heat input)
- Coal cleaning plants (thermal dryers)
- Kraft pulp mills
- Portland cement plants
- Primary zinc smelters
- Iron and steel mill plants
- Primary aluminum ore reduction plants
- Primary copper smelters
- Municipal incinerators (>50 tons/day)
- Acid plants (hydrofluoric, sulfuric, nitric)
- Petroleum refineries
- Lime plants
- Phosphate rock processing plants
- Primary lead smelters
- Fuel conversion plants
- Sintering plants
- Secondary metal production facilities
- Chemical process plants
- Fossil fuel boilers (>250 million BTU/hr input)
- Petroleum storage and transfer facilities (>300,000 barrels storage)
- Taconite ore processing facilities
- Glass fiber processing plants
- Charcoal production facilities

Under the federal rules, all other newly proposed sources are subject to a "major source" threshold of 250 tons/year of air emissions. Many states have followed these federal requirements for threshold determination of new sources, but it is wise to confirm the status of these definitions with your state environmental agency.

If the proposed project is a modification or addition to a ''major stationary source'' in an attainment area, you will be determining the PSD pollutants that need review according to significant emission rate (SER) increases identified in Table 6.1.

If your proposed modification results in a net emissions increase above the SER for any of the listed pollutants, the source is subject to PSD review for that particular pollutant.

Finally, there is a method to avoid PSD review by agreeing to accept enforceable permit limits that restrict emissions to levels below the SER or major source levels for all PSD pollutants. This solution must be reviewed carefully with facility management to ensure that these restrictions will not unduly affect the day-to-day operations of the facility. Facilities will also be required to provide the agency with enough monitoring data to prove that the annual limit is not exceeded. Operators must be thoroughly aware of these limitations to prevent permit violations caused by inadvertently operating at production levels above the allowable range.

Table 6.1 New Source Review (NSR) significant emission rates (SER)

Pollutant	Significant Emission Rate (tons/year)
Sulfur dioxide	40
Particulate matter (TSP)	25
Particulate matter (PM_{10})	15
Nitrogen oxides	40
Carbon monoxide	100
Volatile organic compounds (ozone)	40
Lead	0.6
Sulfuric acid mist	7
Total fluorides	3
Total reduced sulfur (TRS)	10
Reduced sulfur compounds	10
Hydrogen sulfide	10
Asbestos	0.007
Beryllium	0.0004
Mercury	0.1
Vinyl chloride	1
Any regulated pollutant	Class I impact[a]

[a] Any emission rate for a source located within 10 km of a Class I area that causes impacts of 1 Hg/m3, 24-hour average or greater.

6.3 PREAPPLICATION MONITORING

New Source Review applications generally must include a specified amount of preapplication air quality monitoring. These requirements are triggered by lack of specific ambient air quality data for the pollutant you are proposing to increase. Although monitoring may not be technically difficult, the amount of time required to collect the data is often a surprise to the project team. A full year of air quality monitoring may be necessary to satisfy the environmental agency. For most projects, this length of time can create significant delays in the overall project schedule. The amount of preapplication monitoring (if any) should be determined with the environmental agency early in the process, before the project's permitting schedule is developed.

6.4 DETERMINING BEST AVAILABLE CONTROL TECHNOLOGY (BACT)

For those sources subject to PSD review, a determination of the Best Available Control Technology (BACT) will be required. The EPA requires that the BACT analysis be done using the "top-down" approach. This means that the most effective method of controlling a pollutant is analyzed first; then less effective controls are evaluated in descending order.

For each potential control technology, you will need the following information:

- Description of the technology or process,
- Capital cost of installation,
- Quantity of pollutant removed (annually), and
- Cost-effectiveness calculation.

Table 6.2 shows an example of a cost-effectiveness determination. The EPA has developed guidelines for cost-effectiveness calculations, which are used in the example. However, the particular details for your project, including the capitalized interest rate that your company pays, may differ and should be discussed with the lead regulatory agency.

Once the cost-effectiveness calculation is made, the various pollution control technologies are compared. The permit agency and your company will discuss which level of control is BACT, based on precedent and cost factors. Normally, technologies above a certain level of dollars per ton of pollutant removed (e.g., $2,000–$4,000/ton) are not deemed to be cost-effective. But this determination level is constantly evolving upward, so be prepared to

Table 6.2 BACT economic determination example (VOC control by regenerative thermo oxidation)

Design Flow Rate (Scfm)[a]	170,000
Actual Flow Rate	170,000
Equipment Costs	
RCO Purchase and Installation	$4,400,000
Ducting—Purchase and Installation	$1,200,000
Other Equipment	$150,000
TOTAL EQUIPMENT COSTS	$5,750.00
Other Capital Costs	
Site Prep, Foundation, Gas, Electric	$2,500,00
Indirect Capital	$400,000
Capitalized Interest	$465,000
TOTAL CAPITAL COSTS	$9,115,000
Operating Costs	
Annual Supplemental Fuel	$450,000
Annual Electricity	$300,000
Operating Manpower	$20,000
Testing	$20,000
Maintenance (Materials and Labor)	$285,000
ANNUAL OPERATING COSTS	$1,075,000
Miscellaneous Annual Costs	
Heat Exchanger Replacement	$100,000
Catalyst Replacement	$50,000
Taxes/Administration	$100,000
Capital Amortization Costs	
Capital Amortization/Year	$1,400,000
TOTAL ANNUAL COSTS	$2,725,000
Tons VOC Removed Annually	550
Total Cost/TonVOC Removed	$4,955

[a] scfm = standard cubic feet per minute.

compare the project's proposed technology with the technology installed by competitors or other similar installations.

The initial source of information regarding various pollution control technologies is the BACT Clearinghouse administered by the federal EPA. This Clearinghouse summarizes previous BACT determinations across the country. A word of caution regarding the BACT Clearinghouse: Always verify that the technology determined in the Clearinghouse as BACT actually was installed and has achieved its predicted level of pollution control. *Not all projects within the Clearinghouse have actually been built or achieved their emission goals.*

Another possible area of concern is cost. Did the project identified in the Clearinghouse require retrofits or cost more to install than originally estimated? Not every project meets its technical or cost predictions; however, it is up to you, not the agency, to find the answers. The best way to obtain this information is to contact the facility identified in the Clearinghouse and learn more about its actual experience with the technology. Most companies are willing to share technical information on environmental technology so long as it cannot be used to calculate the company's costs of production.

6.5 DETERMINING THE LOWEST ACHIEVABLE EMISSION RATE (LAER)

In essence, the Lowest Achievable Emission Rate (LAER) designates the best pollutant control technology regardless of cost. LAER technology is the required technology if you are constructing a regulated source in a nonattainment area.

LAER may be established through the EPA Clearinghouse or through discussions with manufacturers or design firms familiar with your manufacturing process. As an example of the difference between BACT and LAER, consider NO_x emissions from a fossil fuel boiler. BACT may be defined as good combustion controls, whereas LAER may require the installation of ammonia injection in the flue gas in addition to a good combustion control system.

6.6 MODELING VISIBILITY IMPACTS IN CLASS I AREAS

An area of growing importance in the approval of PSD permits is the analysis of impact to visibility in Class I areas. Class I areas are national parks, wilderness areas, and national monuments designated in the Clean Air Act. If your facility is within 50 kilometers (less in some states) of a Class I area, you will have to analyze the visibility impacts of your project on that area. The local federal land manager will have the authority to review and comment on your visibility analysis and permit application and often will request the facility to mitigate the adverse visibility impacts of the proposed project.

Visibility issues are typically related to NO_x and particulate emissions. There are screening air models and detailed air models to determine the impact of your project. If you are proposing to build a regulated project near a Class I area, you have to understand the basic requirements of visibility analysis and recognize that normal BACT requirements may not be sufficient to avoid visibility impacts.

6.7 PSD AND NSR APPLICATION PROCESS

The application for a PSD or NSR permit is a voluminous document, which includes the following items:

- Permit application form and fee
- Project description
- List of sources
- Emission inventory
- Applicability analysis
- Air modeling information
- Visibility analysis (if required)
- BACT/LAER determinations
- Support appendixes
- Drawings and site map

For a project required to evaluate the emissions of several pollutants and emission points, the application may contain hundreds of pages of written materials, computer printouts, calculations, and drawings. It is extremely important that the documents are clear, complete, and well written. Anything less will increase the likelihood of significant delays in the application review process. The fewer questions, clarification requests, or supplements required, the sooner your permit will be issued.

A key to managing the permit review process is to meet frequently with the permitting agency and to begin these meetings well before the permit application is submitted. Items that should be preliminarily reviewed with the agency include the following:

- An overview of the project
- Potential public concerns
- Air emissions inventory
- PSD pollutant analysis
- Air modeling requirements
- Previous BACT/LAER determinations

Some state agencies also require an Environmental Impact Statement (EIS) or site planning review as part of the New Source Review process. Noise, odor, and other issues regulated by the local or state air authority may also be considered in the permit application process.

After the permitting agency has reviewed and accepted the permit application as complete, you must still respond to questions posed by the agency and provide the agency with any additional information it requests. The failure to provide the agency with requested information can result in the agency's denying your application or, at a minimum, a delay in processing your application.

Once the agency decides to issue the New Source Review permit, it is still required to publish the draft permit determination for public comment. Public comment periods are normally 30 days. At the close of the comment period, the agency must respond to the comments in writing and complete its permit determination. Public concerns can lead to proposed changes in the permit and possible delays in permit issuance. Depending on the public's level of concern, the agency may also require a public hearing.

6.8 IMPLEMENTING PERMIT REQUIREMENTS

A PSD or NSR permit is a legally enforceable document. It is necessary to obtain the permit before beginning construction and to comply with the terms of the permit once it is issued. EPA can issue significant penalties to companies that commence construction of a new source or modification of an existing source prior to receiving a final PSD or NSR permit. Permit compliance will likely require new operator training manuals, new monitoring or testing equipment, and revisions to reporting and data systems. These requirements should be identified and implemented well in advance of the actual start-up of your new facility or modification. The permittee is also typically required to notify the regulatory agency once construction is completed and operation of the project has started.

7

AIR OPERATING PERMITS

Regulation: 40 CFR Parts 70 and 71

For many years, air permits have been required to construct or modify an industrial source of air emissions. In Title V of the 1990 Clean Air Act Amendments, Congress directed EPA to require major emission sources to obtain a permit to operate as well. The intent of this requirement is to bring all major emission sources under a permit program, regardless of when they were constructed, and to consolidate existing air emission permits into one comprehensive document.

The Air Operating Permit (AOP) program is designed to more clearly establish permit limits and methods of determining compliance with those limits. All industrial facilities included in the Title V permit program will be subject to more comprehensive compliance monitoring, record keeping, and certification requirements than previously mandated. The environmental manager very likely will have to review and expand the facility's record-keeping and monitoring systems to meet the requirements of the Air Operating Permit. The facility operations manager also has to be more aware of permit requirements and ongoing compliance issues, inasmuch as the AOP program requires the responsible manager to certify compliance on a periodic basis.

When existing facilities applied for their AOPs, they were required to compare their current operations with their historical air permit applications to ensure that all current operations were properly permitted. As a result of this comparison, facilities sometimes identified production increases or facility modifications that had not obtained all necessary air permits. If compliance issues were identified, the permittee was required to identify those issues in the Title V application and propose a compliance plan to resolve them. Resolution of any compliance issues identified through the Title V process is also a major responsibility of the facility environmental manager.

7.1 QUALIFYING FACILITIES

Under the applicable federal regulations, new or existing industrial facilities are required to obtain an AOP if the source has emissions or potential emissions above the thresholds identified in Table 7.1.

State AOP thresholds can be different from the federal standards and should be reviewed by the environmental manager. Sources in nonattainment areas may also be subject to different criteria. Some state rules will require sources subject to new source performance standards to obtain an AOP even if the total facility emissions are below the qualifying thresholds.

The environmental manager for a facility that is capable of restricting its emissions through process changes or production controls may also want to consider the possibility of exempting the facility from the Title V program by agreeing to restrict potential emissions from the facility to federally enforceable limits below the AOP thresholds. This approach is referred to as becoming a ''synthetic minor'' source and can allow a facility to avoid the costly annual permit fees and environmental management costs relating to the AOP program. The implications of agreeing to restrictions below the AOP thresholds should be discussed thoroughly with the facility manager and operations managers before the facility accepts synthetic minor status, as limitations on facility production may outweigh the benefits of avoiding the AOP requirements.

Table 7.1 General Air Operating Permit (AOP) thresholds

Category	AOP Required If Annual Emissions Exceed
• Any criteria pollutant (e.g. NO_x, CO, etc.)	100 tons
• Any single hazardous air pollutant	10 tons
• Total of all hazardous air pollutants	25 tons

Table 7.2 General Air Operating Permit application elements

1. Facility information—name, address, phone, contacts
2. Process description—narrative and process flow diagrams of the normal mode of operation and production
3. Alternative operating scenarios—similar information for alternative production scenarios that would change emissions
4. Emission source identification
5. Emissions inventory—for each operating scenario
6. Applicable air regulations
7. Proposed *de minimis* or other exemptions
8. Controls
   a. Air pollution equipment
   b. Current limits
   c. Compliance testing and monitoring methods
9. Compliance status summary—including proposed compliance plan and schedule for areas not in compliance
10. Compliance certification—by responsible official (e.g., facility manager)

7.2 APPLICATION PROCESS

If you have determined that your existing facility or new source qualifies for the AOP program, the next step in the process is to prepare an application. Most existing facilities have already completed this process, but any new facility or existing facility that proposes increased emissions above the threshold will be required to apply. The initial application for a Title V Air Operating Permit is normally a complex document, rather than a simple form with a few attachments. An outline of a Title V Air Operating Permit application is shown in Table 7.2.

The application requires an analysis of all air regulation requirements applicable to the facility as well as an extensive legal review that goes well beyond the provisions of existing air permits. The application also requires that the completeness and current compliance be certified by a responsible company official, normally the plant manager.

7.3 IMPLEMENTING AIR OPERATING PERMITS

At this point, experience with implementing AOPs is limited. A stated intent of the AOP program was not to add any additional permit limits. Although this may have been the intent of the law, AOPs usually result in significant increases in monitoring, record keeping, testing, and the use of surrogate parameters to evaluate the operation of a facility. EPA is also encouraging states to transform these surrogate parameters (e.g., opacity as a particulate limit) into enforceable permit requirements. Although normal allowances for start-up, shutdown, and mechanical malfunctions remain in the law, the AOP program has increased significantly the need to document such occurrences at a facility. Without good documentation, the environmental manager may find these incidents classified by the agency as permit violations.

Perhaps the most significant change associated with Title V permits is the need for periodic certification of compliance by the facility manager. Before signing such a certification, most facility managers require a detailed review of the facility's compliance status. The environmental manager is typically the person asked to provide these assurances to the responsible company official.

The key elements of a successful AOP management program are (1) detailed analysis of the permit requirements, (2) establishment of an expanded monitoring and record-keeping system, (3) training for environmental and operating personnel, and (4) periodic reviews for compliance certification. In addition, the environmental manager must develop a system for managing change. Many facilities identify alternative operating scenarios in their permit application, and the AOP establishes different permit conditions for each scenario. It is critical that a facility's compliance monitoring and record-keeping procedures shift to reflect each change in operating scenario. Moreover, the normal industrial facility is always evaluating projects to move product lines, increase production, save energy, or modify the manufacturing process in some way. Some of these changes may require a modification to the AOP. The environmental manager must develop a procedure to document all new emission points, fugitive sources, and process changes to ensure continuous compliance with the provisions of the AOP.

7.4 COMPLIANCE ASSURANCE MONITORING (CAM)

When the Air Operating Permit program was conceived, EPA wanted to improve its ability to determine that a facility subject to the program was in continuous compliance with the air permit limits. Annual source tests and other historical compliance determination methods were deemed inadequate for demonstrating compliance. As part of its AOP program, EPA developed additional monitoring requirements for larger sources of air emissions. The Compliance Assurance Monitoring (CAM) rule was promulgated by EPA in 1997 and took effect for facilities whose AOP applications were not complete in April 1998. A facility whose permit was issued prior to April 1998 is subject to the rule upon renewal of the AOP or when a significant modification request is submitted to EPA. In general, the rule applies to certain major sources whose emissions exceed established thresholds and who use pollution control devices to achieve compliance.

If your facility is subject to the AOP rule, a CAM determination can be made by following the flowchart in Figure 7.1 for each source in your permit (or application):

After you develop the list of sources and pollutants subject to the CAM requirement, the next step is to develop a CAM plan for each unit and pol-

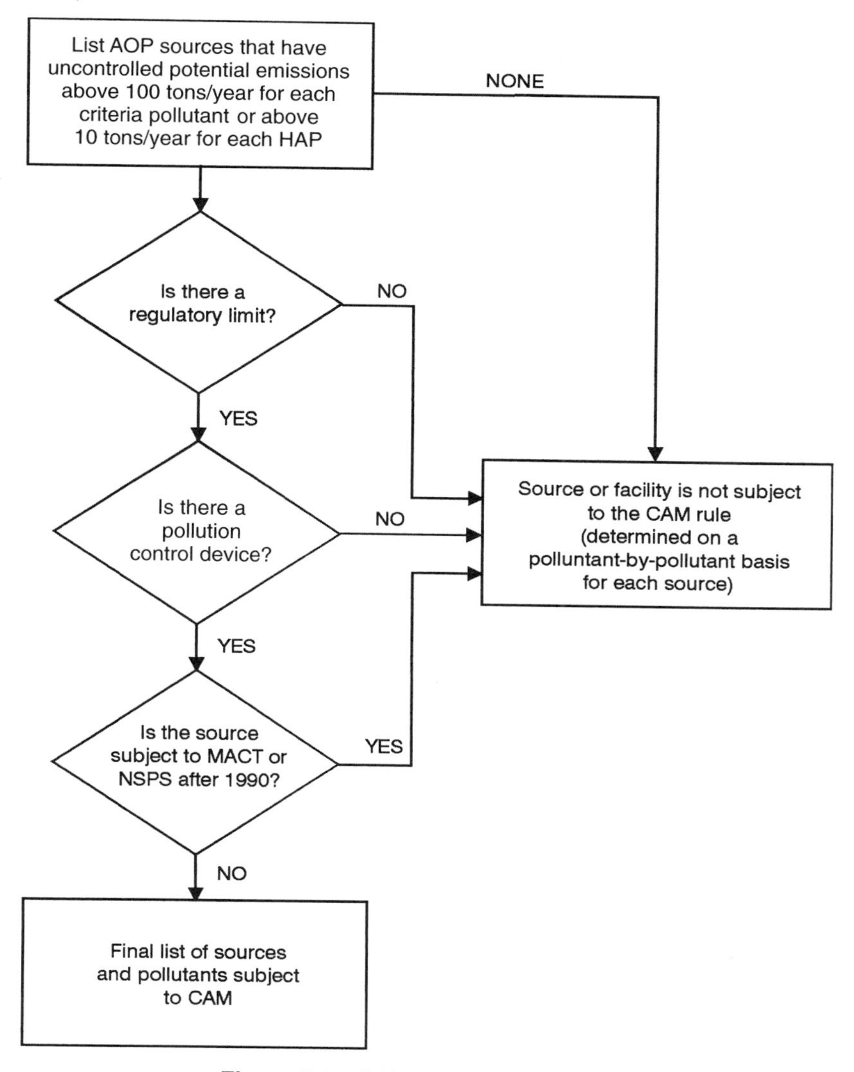

Figure 7.1 CAM determination chart

lutant. You are required to submit the plan to the environmental agency that administers your AOP. The basic elements of the CAM plan include the following:

- Identification of the type of monitoring and indicators for each pollution control device
- Identification of the indicator ranges
- Identification of the performance criteria that demonstrate "normal" operation

Depending on the wording of your permit, once a selected indicator is exceeded, the facility must take prompt corrective action. Failure to act immediately can be viewed as a permit violation. EPA has also tried to persuade some sources to accept these surrogate indicators as actual emission limits. For instance, an electrostatic precipitator's performance parameters may include amperage, voltage, and spark rate readings. Although these parameters indicate a well-functioning precipitator, environmental managers should be reluctant to agree that these parameters are accurate measures of particulate emissions. If a facility is experiencing chronic control device problems, the authorized agency can require the facility to implement a quality improvement plan that includes a formal written plan and schedule for correcting the problems. Because the CAM rule is relatively new, experience with it is limited. It will take several years for EPA and industry to determine how Compliance Assurance Monitoring will be used to measure compliance.

8

NATIONAL EMISSION STANDARDS FOR AIR POLLUTANTS

Regulations:	NSPS—40 CFR Part 60
	MACT—40 CFR Part 63

Management of air issues requires a working knowledge of the technology-based national emission standards for both criteria and hazardous air pollutants. Some specific criteria pollutants relating to particular industries are regulated through the creation of federal New Source Performance Standards (NSPS). The environmental manager should develop a familiarity with the NSPS requirements applicable to his or her facility and industry. These standards may trigger certain pollution control requirements even when no other air permits are required. A relatively new group of national emission standards are being developed for hazardous air pollutants. EPA is in the process of promulgating Maximum Achievable Control Technology (MACT) standards for a long list of hazardous air pollutants (HAPs). Once a MACT standard has been issued for a specific category of sources, the owner of a facility will have three years to comply with the standard.

8.1 NEW SOURCE PERFORMANCE STANDARDS (NSPS) SOURCES

The federal New Source Performance Standards (NSPS) apply to nearly 80 air source categories and include most major industrial sources. NSPS requirements are normally expressed as emission limitations and are specific to process emissions from a particular source. They may also vary by the date a source was constructed. The purpose of the NSPS program is to improve air quality by requiring new plants to install state-of-the-art pollution control. Not all pollutants emitted from a facility or sources of emission are subject to new source standards. For example, in Subpart J of the NSPS (the subpart for petroleum refineries) specific sources, such as the fluid catalytic cracking unit catalyst regenerator, are identified as sources subject to NSPS requirements for particulate, CO, and SO_2. Other refinery sources, such as a Claus sulfur recovery unit, are also subject to NSPS but for a different list of pollutants. The environmental manager should review the specific subpart applicable to the facility and determine which specific emission sources and pollutants are included in the NSPS.

8.1.1 NSPS Triggers

If New Source Performance Standards have been developed for your facility's category of source, the next step is to determine whether the planned activity triggers the application of NSPS. If your facility is constructing a new source of air emissions, NSPS limits will certainly apply. If a source subject to NSPS requirements is being reconstructed or modified, there are two basic factors to evaluate in determining whether NSPS apply. First, if there is an increase in emissions of NSPS pollutants, NSPS will apply. If the reconstruction cost is 50% or greater than the cost of building a new facility, NSPS will also apply, regardless of whether there is an increase in emissions. Evaluation of the reconstruction cost normally involves extensive capital cost estimating and judgments as to what construction activities are within the scope of the modification. If you conclude that NSPS do not apply to your proposed activity, keep your evaluation documents on file, because EPA or your state environmental agency may ask to review your analysis.

8.1.2 NSPS Performance Provisions

New Source Performance Standards also contain detailed monitoring and compliance demonstration provisions. NSPS regulations often require facilities to determine compliance through the use of continuous emission monitors (CEM). For example, the NSPS for lead smelters require continuous monitoring for opacity and sulfur dioxide. NSPS also require facilities to conduct performance tests within a specified time after facility start-up to demonstrate that the facility has met the new source emission limitations.

8.2 NATIONAL EMISSION STANDARDS FOR HAZARDOUS AIR POLLUTANTS (NESHAP)

Early air pollution laws focused on controlling the so-called criteria pollutants. As knowledge of, and concern about, chemical substances that may be health hazards (carcinogenic or otherwise) increased, laws were developed to control the emission of hazardous air pollutants, or "air toxics." The federal standards governing these emissions, known as the National Emissions Standards for Hazardous Air Pollutants (NESHAP), were completely overhauled by Title III of the Clean Air Act Amendments of 1990.

The 1990 Amendments gave EPA authority to develop an entirely new program of air pollution control regulations aimed at significantly reducing the amount of HAPs emitted by various categories of industrial sources. EPA was directed to develop Maximum Achievable Control Technology (MACT) standards for reducing HAP emissions. The MACT category of controls establishes emissions standards based on technology achievement rather than economic considerations. If the MACT available today does not reduce carcinogens to a concentration that yields a one in one million residual risk level (or lower), then EPA may impose even tighter control standards.

The MACT development process is currently under way. EPA has divided industry sources into various subcategories. MACT standards are scheduled to be promulgated by the year 2000, but it appears that EPA may be somewhat behind in achieving this ambitious schedule. Many states have adopted HAPs or air toxics programs in addition to the federal program. State programs typically require screening models at a facility's property boundary to determine whether hazardous air emissions create off-site impacts that should be further evaluated. Further evaluation normally consists of more accurate air dispersion modeling and, possibly, an environmental risk assessment.

Hazardous air pollutants are regulated in other ways by the environmental laws. HAPs must be evaluated in your Air Operating Permit evaluation (Chapter 7) and are considered in the development of Risk Management Plans (RMPs) (Chapter 10). Releases of HAPs may also be governed by the release reporting requirements contained in the Emergency Planning and Community Right-to-Know Law (Chapter 20) and the Superfund Law (Chapter 19). All of these regulatory requirements must be considered by the environmental manager in developing a comprehensive management plan for hazardous air pollutants.

8.2.1 List of Hazardous Air Pollutants

Congress established a list of 189 hazardous air pollutants in Section 112 of the Clean Air Act (CAA) of 1990. The list of HAPs may be revised by EPA to either add or delete substances. When evaluating HAPs at your facility, remember to obtain the most recent list from EPA or your state environmental agency.

8.2.2 MACT Determinations

Maximum Achievable Control Technology is defined in the regulations as "the emission limitation which is not less stringent than the emission limitation achieved in practice by the best controlled similar source, and which reflects the maximum degree of reduction in emissions of HAPs (including prohibition of emissions) that the permitting authority, taking into consideration the cost of achieving such emission reduction and any non-air quality health and environmental impacts and energy requirements, determined achievable by the constructed or reconstructed major source" (40 CFR 63.4). In determining which emission limitation is best, EPA evaluates both existing sources and new sources. For existing sources, the MACT "floor" is determined by the average emission limitation achieved by the best performing 12% of the existing U.S. sources. If fewer than 30 sources exist, the performance of the best 5 sources is averaged. In practice, many sources have limited data or no data on the HAP emissions EPA is required to evaluate. Consequently, industry groups have joined EPA to develop the scientific emissions data required to perform the MACT evaluation. For some HAPs, test methods do not exist, and until they are developed, EPA will not be able to evaluate MACT unless a surrogate pollutant is designated. In the case of new sources, MACT will be equal to the best emission limitation achieved in practice by a similar source. Obviously, this limitation will become more stringent over time as control technology improves.

8.2.3 MACT Implementation

When a MACT standard has been established for a source category, the federal rule calls for implementation within three years. The MACT standard must be included in any Title V Air Operating Permit, which will entail reopening any operating permit that has already been issued. Although MACT determinations do not necessarily dictate specific technology, the emissions limitations are established on the basis of observed performance of existing technology. It is up to the individual facility to review its sources and determine the best technology to achieve the MACT limitations. In some cases process chemistry revisions may be the most economic means to achieve compliance. MACT standards recently promulgated for the pulp and paper industry established three alternatives for meeting MACT in regard to the emissions from the process condensates associated with pulping wood fiber. EPA will also allow facilities to use innovative technology to meet the standards.

In practice, implementing MACT may become a process engineering challenge. Once the MACT standard is promulgated and published in the *Federal Register*, affected facilities have only three years to meet the standard, which is hardly enough time to plan, permit (if necessary), and implement a major construction project.

8.2.4 Early Reduction Extensions

The NESHAP rules also provide incentives to implement HAP reductions early. If your facility achieves a 90% reduction in HAPs (95% for particulate HAPs) from its 1987 level of emissions before the MACT standard is promulgated for the facility, the facility can apply for an extension of the MACT compliance deadline. Once a facility fulfills the extension program requirements, it can be granted an additional six years to meet the MACT standards.

9

OZONE-DEPLETING CHEMICALS (ODCs)

Regulation: 40 CFR Part 82

During the 1980s, concern for the depletion of the earth's ozone layer increased. Many scientists attributed the depletion to the release of man-made chlorofluorocarbons (CFCs) and other compounds. These substances were commonly produced and used in mechanical devices such as air conditioners, fire suppression systems, aerosol products, and refrigeration systems. An international agreement was reached to ban the future production of many of these substances and phase out their current use. This agreement, known as the Montreal Protocol, was signed by the United States and other countries in 1987. In response to the Montreal Protocol, Congress enacted Title VI of the 1990 Clean Air Act Amendments to provide EPA authority to regulate ozone-depleting chemicals (ODCs). EPA, in turn, developed regulations regarding ODCs, which became effective in 1995.

In the case of existing operations utilizing such chemicals, the law imposes practical measures to ensure that the existing inventories of these substances are contained, recycled, or destroyed. Environmental managers whose facilities still use such substances must develop a program to ensure compliance with these regulatory requirements.

9.1 SUBSTANCES OF CONCERN

In conjunction with the international protocol, Congress listed ozone-depleting substances in classes based on their depletion potential. A sub-

Table 9.1 Ozone depletion potentials of various substances[a]

Substance	Ozone-Depletion Potential
CLASS I	
chlorofluorocarbon-11 (CFC-11)	1.0
chlorofluorocarbon-12 (CFC-12)	1.0
chlorofluorocarbon-13 (CFC-13)	1.0
chlorofluorocarbon-111 (CFC-111)	1.0
chlorofluorocarbon-112 (CFC-112)	1.0
chlorofluorocarbon-113 (CFC-113)	0.8
chlorofluorocarbon-114 (CFC-114)	1.0
chlorofluorocarbon-115 (CFC-115)	0.6
chlorofluorocarbon-211 (CFC-211)	1.0
chlorofluorocarbon-212 (CFC-212)	1.0
chlorofluorocarbon-213 (CFC-213)	1.0
chlorofluorocarbon-214 (CFC-214)	1.0
chlorofluorocarbon-215 (CFC-215)	1.0
chlorofluorocarbon-216 (CFC-216)	1.0
chlorofluorocarbon-217 (CFC-217)	1.0
halon-1211	3.0
halon-1301	10.0
halon-2402	6.0
carbon tetrachloride	1.1
methyl chloroform	0.1
CLASS II	
hydrochlorofluorocarbon-22 (HCFC-22)	0.05
hydrochlorofluorocarbon-123 (HCFC-123)	0.02
hydrochlorofluorocarbon-124 (HCFC-124)	0.02
hydrochlorofluorocarbon-141(b) (HCFC-141(b))	0.1
hydrochlorofluorocarbon-142(b) (HCFC-142(b))	0.06

[a] The environmental manager should refer to 40 CFR 82, Subpart (A), Appendixes A and B, for a comprehensive list of all Class I and Class II regulated substances.

stance's potential to deplete the ozone is established by comparing it with the potential of chlorofluorocarbon-11 (CFC-11). In general, CFCs, halons, carbon tetrachloride, and methyl chloroform make up the Class I category, or highest-priority substances. Class II substances consist solely of hydrochlorofluorocarbons (HCFCs). The initial ozone-depletion potentials of substances known at the time of the 1990 Amendments were listed in Title VI of the Clean Air Act, as shown in Table 9.1.

9.2 PRODUCTION AND CONSUMPTION PHASEOUTS

Industrial facilities producing listed ozone-depleting substances are subject to a production phaseout program established under Title VI. To industry's

Table 9.2 Class I substance phaseout schedule

Date	CFCs, Halons, and Other Class I Substances	Carbon Tetrachloride	Methyl Chloroform
1991	85%	100%	100%
1992	80%	90%	100%
1993	75%	80%	90%
1994	65%	70%	85%
1995	50%	15%	70%
1996	40%	15%	50%
1997	15%	15%	50%
1998	15%	15%	50%
1999	15%	15%	50%
2000	0%	0%	20%
2001	0%	0%	20%
2002	0%	0%	0%

credit, this phaseout schedule has been voluntarily exceeded in many instances. Facilities producing Class I substances are subject to the phaseout deadlines shown in Table 9.2. As noted in the table, all substances in Class I except methyl chloroform must be phased out of production by the year 2000. Methyl chloroform production is banned effective January 1, 2002.

There are some exceptions to the phaseout deadline for Class I substances, such as methyl chloroform, that are used in medical devices or for airplane metal fatigue testing. There also are exceptions for some halogens used for fire protection and exports of halogens to developing countries that have agreed to use the substances properly under the provisions of the Montreal Protocol. The environmental manager should refer to the regulations for a detailed explanation of these exemptions.

Class II substances are subject to a longer phaseout and use period. In general, the HCFC substances in this category will not be subject to phaseout restrictions until 2015. At that time, manufacturers will be required to hold production to 1986 levels. By the year 2030, production of HCFC substances must cease. New uses of HCFC are not lawful after January 1, 2015. Class II substances are granted limited exemptions for use in medical devices or as exports to developing nations.

9.3 MANAGING REMAINING ODC-CONTAINING EQUIPMENT

Industrial facilities that are not involved with manufacturing ODCs may be affected by the ODC regulations if they engage in servicing motor vehicle air conditioners and other equipment that contains ODCs or labeling products containing or manufactured with ODC substances. As time goes on, these requirements will become less onerous as the regulated ODCs are phased out of use and older vehicles and equipment are replaced.

Beginning with the 1993–1994 models, motor vehicle manufacturers began to convert air-conditioning systems from R-12 (CFC-12) to other refrigerants. If you are still servicing motor vehicles that contain the regulated R-12 coolant, the technician performing the service must be trained and certified. Specifically approved equipment must be used to capture and contain any removed refrigerants. If the refrigerant is captured properly and reused in the vehicle, the procedure meets the requirements of regulation. If the reclaimed material is sent off-site for recycling or disposal, records must be kept of the quantity of material reclaimed and its ultimate destination. From a practical standpoint, many industrial facilities choose to send their motor vehicles to off-site service shops for cooling system maintenance rather than develop their own programs to meet the training, certification, and other regulatory requirements.

A similar regulatory program applies to ODC systems used in nonvehicle mechanical equipment such as air conditioners and refrigeration systems. These types of refrigeration systems are more likely to be used for a period of years. Under the regulations, a certified technician with the proper evacuation equipment must be engaged to service these ODC-containing systems. The question for most facilities is whether it is more cost-effective to provide training and certification of internal maintenance personnel or to retain certified outside contractors to service the equipment. Another management option is to convert the equipment to other coolants, which typically requires an engineering study to determine the cost of such a conversion and whether the conversion will reduce the cooling capacity of the equipment.

The ODC rules also establish a labeling program regarding products that contain or, in some cases, are manufactured with ODCs. If this requirement applies to a product manufactured at your facility, the following label must be applied to the container or product.

> Warning: Contains (or manufactured with, if applicable) [insert name of substance], which harms public health and environment by destroying ozone in the upper atmosphere.

The label is required to be "clearly legible and conspicuous." Exceptions to this labeling requirement apply to trace residues, research and development activities, and repairs. Specifics of the label design and exemptions are detailed in the rule.

In 1998, EPA issued a final rule governing the manufacture, release, use, and disposal of halon and halon-containing equipment. Halon is widely used in fire-extinguishing equipment. The final rule establishes detailed standards governing the proper disposal of halon and halon-containing equipment and limiting the release of halon during training in the use of, and testing, repair, and disposal of, halon-containing equipment. The rule also prohibits the manufacture of halon, subject to certain exceptions.

10

RISK MANAGEMENT PLANS

Regulation: 40 CFR Part 68

Several laws have been established whose objective is to protect the general public in the vicinity of an industrial facility from toxic chemical releases. In the 1990 Clean Air Act Amendments, Congress established a requirement to develop risk management programs designed to protect the public from potential catastrophic air releases of certain toxic chemicals and flammable substances. In 1996, the Environmental Protection Agency (EPA) promulgated a rule that requires facilities to determine whether they are subject to the risk management program rule and, if so, to develop by June 21, 1999, both a program and a risk management plan (RMP) to handle accidental releases of 140 regulated substances. The requirements of this rule are not the same for all facilities and will vary according to the proximity of public receptors, the safety record of the facility, and other relevant factors.

A facility can reduce its exposure to RMP requirements by reducing quantities of chemicals, eliminating on-site storage or use of chemicals, or redesigning its manufacturing processes to accomplish the aforementioned goals. If a facility is required to develop an RMP, the most significant challenge for the environmental manager will likely be the development of a public relations and communication plan to inform the surrounding community of the potential risks associated with the facility. There is no easy way to communicate to the public a ''worst-case'' disaster scenario relating to the catastrophic failure of a large propane tank or a vessel containing toxic chemicals. Fortunately, industrial safety records in the United States are excellent, and the probability of such an occurrence is low.

10.1 RMP APPLICABILITY

The RMP requirement generally applies to an industrial facility classified as a stationary source that handles RMP-regulated substances. When RMP substances are being transported, they are regulated by the Department of Transportation and exempt from the RMP rule. However, once those transported substances are handled by on-site personnel or connected to the stationary source, they have to be considered in the applicability determination. For instance, a rail car containing ammonia used in a lubrication oil refining process is exempt from the RMP rule while in transport. It becomes part of the stationary source and subject to RMP requirement once it is connected to the process or turned over to the facility for storage and handling.

The next step in determining the applicability of the RMP rule is to review

Table 10.1a Example list of regulated toxic substances

Chemical Name[a]	CAS No.[b]	Threshold Quantity (lb)
Ammonia (anhydrous)	7664-41-8	10,000
Ammonia (conc 20 wt % or greater)	7664-41-7	20,000
Chlorine	7782-50-5	2,500
Chlorine dioxide [chlorine oxide (ClO_2)]	10049-04-4	1,000
Formaldehyde (solution)	50-00-0	15,000
Hydrochloric acid (conc 37 wt % or greater)	7647-01-1	15,000
Hydrogen fluoride/hydrofluoric acid (conc 50 wt % or greater)	7664-39-3	1,000
Methyl chloride [methane, chloro-]	74-87-3	10,000
Methyl isocyanate [methane, isocyanato-]	624-83-9	10,000
Nitric acid (conc 80 wt % or greater)	7697-37-2	15,000
Oleum (fuming, sulfuric acid) [sulfuric acid, mixture with sulfur trioxide][c]	8014-95-7	10,000
Sulfur dioxide (anhydrous)	7446-09-5	5,000
Toluene 2, 4-diisocyanate [benzene, 2, 4-diiasocyanato-1-methyl-][c]	584-84-9	10,000
Vinyl acetate monomen [acetic acid ethenyl ester]	108-05-4	15,000

[a] conc = concentration.

[b] Chemical Abstract Service Number.

[c] The mixture exemption does not apply to the substance.

the list of regulated substances identified in the rule. This list includes both toxic chemicals (77) and flammable substances (63). Representative chemicals and flammables from the list and the associated threshold determination levels are shown in Tables 10.1a and 10.1b.

To determine whether your facility exceeds the applicable threshold quantity, it will be necessary to inventory each regulated substance on-site. The inventory of each substance is defined as the largest amount or peak quantity stored on-site at any given time. The environmental manager will need to work closely with operations management, maintenance personnel, and purchasing to verify these quantities. Remember to consider all chemicals used at the facility, including maintenance chemicals such as acids used for equipment cleaning. The total inventory for each regulated substance is the sum of the quantities identified on the entire site. Some important exemptions to the inventory are available and may impact the inventory of a substance. The exemptions are identified in the threshold determination section of the rules.

The complete list of exemptions for the threshold determination are summarized in Table 10.2.

Once the inventory of regulated substances is complete, the environmental manager must determine whether the facility exceeds the maximum annual threshold levels. The manager also should consider whether the amounts of regulated substances normally manufactured, used, or stored by the facility can be reduced below the threshold levels, which may allow the facility to avoid the requirements of the RMP rule. For example, does the facility really need a propane storage tank exceeding 10,000 pounds. (approximately 2,380 gallons)? For certain facilities the RMP requirement can be avoided simply by installing a smaller tank. Once the list of substances exceeding the thresholds is finalized, the next task is to determine which RMP program level applies to the facility.

Table 10.1b Example list of regulated flammable substances

Chemical Name	CAS No. [a]	Threshold Quantity (lb)
Acetaldehyde	74-86-2	10,000
Butane	106-97-8	10,000
Ethane	74-84-0	10,000
Ethyl acetylene [1-butyne]	107-00-6	10,000
Hydrogen	1333-74-0	10,000
Methane	74-82-8	10,000
Propane	74-98-6	10,000
Vinyl chloride [ethene, chloro-]	75-01-4	10,000

[a] Chemical Abstract Service Number

Table 10.2 List of RMP threshold determination exemptions

Regulated Substances Contained in or Used in
Certain mixtures
Explosives, under certain provisions
Articles
Structural components
Products for routine janitorial maintenance
Employees food, drugs, cosmetics, or other personal items
Process water or noncontact cooling water
Air used as compressed air or combustion air
Laboratory substances, except for specialty chemical production, pilot scale operations, or activities outside the lab

10.2 PROGRAM LEVEL DETERMINATION

The environmental manager's evaluation of compliance requirements will be done on a process basis for those substances regulated by the RMP rule and exceeding the threshold quantities. Facilities may have to develop a program for as many as three different compliance levels. To determine the compliance levels that are applicable to a particular facility, the facility must first identify which of its processes have to be evaluated. The definition of *process* under the RMP rule is as follows:

Process. Any activity involving a regulated substance including any use, storage, manufacturing, handling, or on-site movement of such substances, or a combination of such activities. For purposes of this definition, any group of vessels that are interconnected or separate vessels that are

Table 10.3 General description—RMP program levels

Program Level 1. A regulated process with no record, within the last five years, of a release of a regulated substance causing off-site impacts and with no current potential to impact any off-site public receptor. To qualify for this level, the facility must have implemented an emergency response plan.

Program Level 2. A regulated process that does not meet Program Level 1 requirements but is not subject to the requirements of Program Level 3.

Program Level 3. A regulated process that does not meet Program Level 1 requirements and is subject to either of the following criteria:

1. The process is in SIC[a] code 2611, 2612, 2819, 2821, 2865, 2869, 2873, 2879, or 2911.
2. The process is subject to OSHA[b] Process Safety Management (PSM) standard, published at 29 CFR § 1910.119.

[a] Standard Industrial Classification
[b] Occupational Safety and Health Act

located such that a regulated substance could be involved in a potential release shall be considered a single process (40 CFR 68.3).

Once the various processes have been identified, the next step is to determine which program level applies to each process. The three program levels are described generally in Table 10.3.

In regard to Program Level 3 eligibility, if a process meets the SIC codes or OSHA PSM requirements specified in Program Level 3, it still can qualify for Program Level 1 if it meets the Level 1 requirements.

10.3 PROGRAM LEVEL REQUIREMENTS

The environmental manager may find that some of the required elements of the applicable RMP program level have already been met at the facility as a result of efforts to comply with the OSHA PSM standard. Many industrial facilities also have already developed extensive emergency response plans and coordinated those plans with the local authority responsible for emergency responses. The individual elements of the various RMP program levels are summarized in Table 10.4.

The key distinction between Program Level 2 and Level 3 is the different prevention plan requirements. The Level 3 prevention program requires more

Table 10.4 RMP program level requirements

Program Level 1—Submit an RMP in which you:
- Perform worst-case release scenario analysis, and document that the nearest public receptor is beyond the distance not effected by a toxic or flammable end point.
- Complete a five-year accident history.
- Ensure that response planning is coordinated with local emergency planning and response agencies.
- Provide certification of the findings.

Program Level 2—Submit an RMP in which you:
- Establish and implement a management system.
- Conduct a hazard assessment that includes the worst-case analysis.
- Implement Program Level 2 prevention steps.
- Develop and implement an emergency response program.
- Submit data on prevention program elements.

Program Level 3—Submit an RMP in which you:
- Establish and implement a management system.
- Conduct a hazard assessment including analysis of worst-case releases and previous five-year accident history.
- Implement Program Level 3 prevention steps.
- Develop and implement an emergency response program.
- Submit data for Program Level 3 prevention processes.

extensive information in the areas of process safety management, hazard analysis, and operating procedures. Level 3 also requires the development of a mechanical integrity program, written procedures for changes, prestart-up safety reviews for new or modified sources, involvement of employees, hot work permits, and contractor evaluations.

10.4 HAZARD ASSESSMENTS

Any industrial facility subject to the RMP rule is required to conduct a hazard assessment for each process, regardless of program level. In the event a process falls under Program Level 1, the hazard assessment is limited to the five-year worst-case release scenario. For processes that are subject to Program Level 2 or 3, several additional requirements are imposed. This hazard assessment is complicated and very technical in nature. Environmental managers will likely have to review the specifics of the rule and consult with experts in such assessments to ensure compliance with these requirements.

One of the most technical aspects of hazard assessment is the air modeling of gaseous releases or flammable explosions. Although EPA has provided to the regulated community information regarding worst-case release distances and ''off-site consequence analysis compliance,'' much of this information is quite conservative in nature and overestimates off-site impacts. Some companies have decided to use this conservative EPA information to avoid criticism that they did not, in fact, do a worst-case analysis, and other companies will likely conduct their own air dispersion analysis with actual data in hope that it will demonstrate fewer off-site consequences. Each facility will have to make its own strategic decision as to how to conduct this assessment. Either approach to modeling is acceptable.

The hazard assessment requirement to conduct worst-case release scenarios is more than a technical management challenge. Explaining the possibility of such an event to the persons potentially impacted by such a release may be the biggest challenge of the entire RMP rule.

10.4.1 Worst-Case Release Scenario

The worst-case release analysis is aimed at determining the greatest distance toxic chemicals or flammables released from the facility would travel. EPA has defined this scenario as the release of the largest quantity of a regulated substance caused by a vessel or process line failure that results in the release traveling the greatest distance from the facility to a public receptor. The environmental manager will have to locate these public receptors in the vicinity of the industrial site on an appropriate map. The RMP rule definition of *public receptors* is as follows:

Public Receptors. Off-site residences, institutions (e.g., schools, hospitals), industrial, commercial, and office buildings, parks, and recreational areas inhabited or occupied by the public at any time without restriction by the stationary source where members of the public could be exposed to toxic concentrations, radiant heat, or overpressure as a result of an accidental release (see 40 CFR § 68.3).

Most facilities will have to conduct some research to identify their worst-case scenario, unless it is an obvious situation such as the largest storage tank for a regulated substance being located at the property boundary. In any event, the environmental manager should document the worst-case determination and any assumptions used in the analysis for purposes of explaining later the facility's rationale to agency inspectors.

In evaluating a Program Level 1 process, only one worst-case release scenario is required to confirm that a release would not produce any off-site impacts. In practice, several analyses may be required to establish the actual worst-case scenario. A Program Level 2 or Level 3 process requires more work. For those program levels, the environmental manager must perform a worst-case release scenario for each regulated toxic and flammable substance at the facility. Additional cases must be included in the evaluation if either the toxic case or the flammable case could impact different public receptors, which is possible when, for example, prevailing wind directions change. Again, the identification of actual worst-case scenarios will likely require multiple cases to be analyzed.

The environmental manager should document all of the parameters and assumptions used in developing each case. Refer to Subpart B of the regulations, "Hazard Assessment," for a full list of the technical requirements.

10.4.2 Alternative Release Scenarios

For processes within the Program Level 2 and 3 categories, the facility will need to develop "alternative release scenarios," which are defined as "more likely to occur" than the worst-case scenario. For instance, instead of a full tank failure, an alternative release scenario might be a rupture of the tank fill line, which throttles the escape rate of the substance and reduces its atmospheric concentration. Possible alternative release scenarios identified by EPA are summarized in Table 10.5.

Table 10.5 Possible alternative release scenarios

- Transfer hose releases caused by splits or sudden hose uncoupling
- Process piping releases from flanges, etc.
- Process vessel or pump releases because of cracks, seal failures, etc.
- Vessel overfilling and spill
- Shipping container mishandling and breakage

10.4.3 Five-Year Accident History

The other requirement of the hazard assessment is the five-year accidental release history for all regulated processes. The environmental manager will have to include in this history all facility incidents that resulted in deaths, injuries, or significant property damage on-site, and deaths, injuries, evacuations, sheltering, property damage, or environmental damage outside the facility. The environmental manager is required to include in the report pertinent data for each case. EPA requires that the data listed in Table 10.6 be reported for each listed accident if the information is known.

10.5 PREVENTION PROGRAM REQUIREMENTS

The environmental manager will have to establish a prevention program for all Program Level 2 and Level 3 processes. Under the provisions of the chemical accident prevention regulations, Program Level 3 facilities are required to prepare a prevention program that includes the following elements:

- Process safety information
- Process hazard analysis
- Operating procedures
- Training
- Mechanical integrity
- Prestart-up review
- Compliance audits
- Incident investigation
- Employee participation
- Hot work permit procedures
- Contractor safety performance programs

Table 10.6 Accidental release report

- Date, time, and approximate duration of the release
- Chemical(s) released
- Estimated quantity released (lb)
- Type of release event and its source
- Weather conditions
- On-site impacts
- Known off-site impacts
- Initiating event and contributing factors
- Whether off-site responders were notified

The Level 2 prevention program, which is less detailed than that for Level 3, establishes procedural requirements for safety information, hazard review, operating procedures, training, maintenance, compliance audits, and incident investigation.

10.6 EMERGENCY RESPONSE PROGRAM

Facilities subject to the RMP rule must coordinate emergency responses with their local emergency response organizations. It is likely that your facility already has some elements of an emergency response program in place because of other environmental or safety regulations. If your facility elects to have employees participate in its emergency response team, then a written emergency response plan is required. The requirements for the written plan are consistent with other emergency response regulations and are outlined in Subpart E of the rule.

10.7 RISK MANAGEMENT PLANS AND OTHER REQUIREMENTS

An RMP is required for all facilities subject to the rule. The elements to be included are detailed in Subpart G of the rule and are summarized as follows:

Risk Management Plan Outline

- Executive summary
- Registration form
- Worst-case release analysis
- Five-year accident history

- Prevention programs
- Emergency response program
- Certifications

The plan must be updated every five years or as required (within six months) for process changes and program level changes. The plan must be available for public review.

Several other RMP requirements should be noted. First, an organizational chart of responsible managers must be developed. This chart should identify the RMP manager, environmental manager, prevention manager, emergency response manager, maintenance manager, operations supervisor, and other pertinent staff, such as the safety manager. The authorized agency is also required to audit the facility's Risk Management Program. A facility can be exempted from these audits if it attains a ''Star of Merit'' ranking under OSHA's voluntary protection program.

11

ASBESTOS MANAGEMENT

Regulations:	40 CFR Part 61, Subpart M (Air)
	29 CFR § 1910.1001 and § 1926.58–(OSHA)

Asbestos is a general term used to describe a number of fibrous mineral substances that are resistant to heat and corrosive chemicals. Asbestos has been used around the world for decades as a fireproofing material in residential, commercial, and industrial buildings. When crumbled or pulverized, asbestos is reduced to a dust of microscopic fibers. These fibers can easily penetrate body tissues if inhaled or ingested and can cause various forms of cancer and asbestosis, a chronic lung disease that can be fatal.

Asbestos materials become friable as they age, when they are exposed to the atmosphere, or when they are damaged by heat or water. The fibrous or fluffy spray-applied asbestos materials used for fireproofing and insulation in many older buildings are often friable. Pipe and boiler insulation materials also often contain asbestos, as do some ceiling tiles, roofing materials, and older floor tiles.

Various federal and state agencies currently regulate the use and removal of fibrous asbestos material. Our focus in this chapter is on the Clean Air Act (CAA) requirements administered by the federal Environmental Protection Agency (EPA) and the worker safety provisions of the federal Occupational Safety and Health Agency (OSHA), because these agencies administer the programs that are most relevant to industrial environmental managers. An overview of these regulatory programs follows.

11.1 CLEAN AIR ACT REQUIREMENTS

Asbestos is regulated as a hazardous air pollutant (HAP) under Section 112 of the Clean Air Act. The asbestos regulations establish standards for asbestos mills and manufacturing facilities as well as standards governing demolition and renovation projects.

The requirements for demolition and renovation projects published at 40 CFR § 61.145 are based on the amount of regulated asbestos-containing material (RACM) that is to be removed. In general, RACM is defined to include asbestos material that may become friable during demolition or renovation activities. The regulations apply if the combined amount of RACM to be stripped, removed, dislodged, cut, drilled, or similarly disturbed is at least 80 linear meters (260 linear feet) of pipe or at least 15 square meters of material (160 square feet) on other facility components, or at least one cubic meter (35 cubic feet) of material on facility components where the length or area cannot be measured. This section of the regulations also requires the owner or operator of a demolition or renovation activity to thoroughly inspect the affected facility where the demolition or renovation operation will occur, for the presence of asbestos, prior to the commencement of any demolition or renovation activity.

11.1.1 Notification Requirements

The regulations establish detailed requirements for notifying EPA of any regulated asbestos removal activities prior to the commencement of such activities. The Notification of Demolition and Renovation form is published at 40 CFR § 61.145 and is included in Appendix H of this book for your review and use.

11.1.2 Procedures for Asbestos Emission Control

These regulations generally provide that RACM should be removed before any activity begins that would break up, dislodge, or similarly disturb the material. When RACM is taken out of a facility in units or in sections, exposed surfaces must be kept adequately wet during cutting or disjoining operations. The material must then be carefully lowered to the floor and to ground level without dropping, throwing, sliding, or otherwise disturbing the asbestos. When RACM is stripped from a facility, it also must be adequately wet during the stripping operation.

Once RACM is removed from a facility, it must be kept adequately wet until collected and contained in leak-tight wrapping or treated in preparation for disposal. The regulations further provide that no RACM shall be stripped, removed, or otherwise handled or disturbed at a facility unless one on-site representative, such as a foreman or management-level person, is trained in the provisions of the Clean Air Act regulations and understands how to comply with the rules.

11.1.3 Preparation of Asbestos for Disposal

Waste disposal of asbestos from demolition and renovation activities must follow a detailed set of standards. In general, asbestos waste from demolition and renovation activities must be placed in leak-tight containers prior to disposal. The containers must be labeled with warning labels that meet requirements specified by OSHA. The regulations also specify labeling requirements for asbestos waste when it is shipped from the removal area to an asbestos disposal site. Records of asbestos waste shipments must be maintained by the owner or operator of the removal activity.

11.1.4 Asbestos Waste Disposal Sites

EPA rules generally require the owner or operator of an asbestos waste disposal site to ensure that there is no discharge of visible emissions to the outside air during the collection, processing, packaging, or transport of RACM and that the asbestos waste must be deposited as soon as practicable at the waste disposal site. There also are detailed requirements regarding the need to mark vehicles used to transport RACM and to complete and maintain waste shipment records. The operator of an active asbestos waste disposal site must limit access to the site by either a natural barrier or warning signs. The site operator also maintains records of the location, depth, area, and quantity of asbestos-containing waste material within the disposal site.

The EPA asbestos regulations provide a useful table that cross-references other federal asbestos regulations. This cross-reference table is reproduced in Table 11.1.

The Clean Air Act authorizes EPA to delegate to each state the authority to implement the asbestos emission control program. Many states have established an all-inclusive asbestos abatement control program and, in some instances, have enacted rules that go beyond the federal requirements.

11.2 OSHA AND WORKER SAFETY

OSHA has promulgated comprehensive requirements regarding worker protection and the removal of asbestos material. The requirements are focused on worker safety in two areas: the first standard applies in all workplaces, and the second applies only to construction activities.

The general industry standard regulating asbestos applies to all occupa-

Table 11.1 Cross-reference to other asbestos regulations, 40 CFR § 61.156

Agency	CFR Citation	Comment
EPA	40 CFR Part 763, Subpart E	Requires schools to inspect for asbestos and implement response actions and submit asbestos management plans to States. Specifies use of credited inspectors, air sampling methods, and waste disposal procedures.
	40 CFR Part 427	Effluent standards for asbestos manufacturing source categories.
	40 CFR Part 763, Subpart G	Protects public employees performing asbestos abatement work in states not covered by OSHA asbestos standard.
OSHA	29 CFR § 1910.1001	Worker protection measures—engineering controls, worker training, labeling, respiratory protection, bagging of waste, 0.2 f/cc permissible exposure level.
	29 CFR § 1926.58	Worker protection measures for all construction work involving asbestos, including demolition and renovation work practices, worker training, bagging of waste, and permissible exposure level.
Mine Safety and Health Administration (MSHA)	30 Part CFR 56, Subpart D	Specifies exposures limits, engineering controls, and respiratory protection measures for workers in surface mines.
	30 CFR Part 57, Subpart D	Specifies exposure limits, engineering controls, and respiratory protection measures for workers in underground mines.
Department of Transportation (DOT)	49 CFR Parts 171 and 172	Regulates the transportation of asbestos-containing waste material. Requires waste containment and shipping papers.

tional exposures to asbestos in the workplace. The regulations establish a permissible exposure limit (PEL) for airborne asbestos. Although there are no laws that require the removal of asbestos from commercial or industrial facilities, all employers must ensure that their employees are not exposed to asbestos material in levels above the applicable PEL. The OSHA regulations specify detailed and often expensive techniques for protecting workers during the course of asbestos abatement projects. The general industry and construction industry standards for asbestos are summarized in the following paragraphs.

11.2.1 Exposure Monitoring

Determinations of employee exposure are based on breathing zone air samples that are representative of the eight-hour Time-Weighted Average (TWA) limit and 30-minute short-term exposures for each employee.

In general industry, employers must do initial monitoring for workers who may be exposed above the PEL "action level."

In construction, daily monitoring must be continued until exposure drops below the PEL action level.

11.2.2 Methods of Compliance

In both general industry and construction, employers must control exposures, using engineering controls to the extent feasible. Where engineering controls are not feasible to meet the exposure limit, they must be used to reduce employee exposures to the lowest levels attainable and must be supplemented by the use of respiratory protection.

11.2.3 Respirators

In general industry and construction, the level of exposure determines the type of respirator that is required. The standards specify the respirator to be used.

11.2.4 Regulated Areas

In general industry and construction, regulated areas must be established where the PELs for airborne asbestos exceed the prescribed permissible exposure limits. Only authorized persons wearing appropriate respirators can enter a regulated area. Warning signs must be displayed at each regulated area and must be posted at all approaches to regulated areas.

11.2.5 Labels

Caution labels must be placed on all raw materials, mixtures, scrap, waste, debris, and other products containing asbestos fibers.

11.2.6 Record Keeping

The employer must keep an accurate record of all measurements taken to monitor employee exposure to asbestos. This record must be kept for 30 years.

11.2.7 Protective Clothing

For any employee exposed to airborne concentrations of asbestos that exceed the PEL, the employer must provide and require the use of protective clothing such as coveralls or similar full-body clothing, head coverings, gloves, and foot covering. Wherever the possibility of eye irritation exists, face shields, vented goggles, or other appropriate protective equipment must be provided and worn.

In construction there are special regulated area requirements for asbestos removal, renovation, and demolition operations. These provisions include a negative pressure area, decontamination procedures for workers, and a ''competent person'' with the authority to identify and control asbestos hazards. The standard includes an exemption from the negative pressure enclosure requirements for certain small-scale, short-duration operations, provided special work practices described in an appendix to the standard are followed.

11.2.8 Hygiene Facilities and Practices

Clean change rooms must be furnished by employers for employees who work in areas where exposure is above the PEL. Lockers must be provided and separated to prevent contamination of the employee's street clothes from protective work clothing and equipment. Showers must be furnished so that employees may shower at the end of the work shift. Employees must enter and exit the regulated area through the decontamination area.

The equipment room must be supplied with impermeable, labeled bags and containers for the containment and disposal of contaminated protective clothing and equipment.

11.2.9 Medical Exams

In general industry, employees must have a preplacement physical examination before being assigned to an occupation exposed to airborne concentrations of asbestos at or above the PEL action level.

In construction, examinations must be made available annually for workers exposed above the PEL for 30 or more days per year or who are required to wear negative pressure respirators.

11.3 ASBESTOS MANAGEMENT PROCEDURES

Asbestos is a highly regulated material. In determining the presence and condition of asbestos-containing materials within your facility and how best to manage those materials, it is critical to engage a competent asbestos management advisor. Determining how to address the dangers posed by asbestos-contaminated materials is an extremely complicated matter. Improperly performed abatement actions often make matters far worse rather than better. Using expert help will make it more likely that your facility will select the proper course of action and that all activities will be performed in accordance with the applicable regulations. An asbestos management program should, in most instances, include the elements discussed in the following paragraphs.

11.3.1 Asbestos Survey

An asbestos survey is conducted for the purpose of locating, identifying, and assessing the condition of asbestos-containing materials in a particular facility. If asbestos-containing materials are found within the facility, facility managers should then consider the wisdom of implementing an asbestos operation and maintenance program.

11.3.2 Asbestos Operations and Maintenance (O&M) Program

The presence of asbestos-containing materials in a building and the resulting potential for occupant exposure does not necessarily mean that removal of the materials is the only course of action. Frequently, potential exposure to asbestos can be controlled with an effective Operations and Maintenance (O&M) program. O&M programs control employee exposure to asbestos material and are designed to prevent disturbance of these materials.

11.3.3 Asbestos Abatement Activities

Facilities with asbestos materials must periodically undertake asbestos abatement activities because of renovation activities, demolition of a building, or the poor condition of the asbestos-containing material. Facility managers must develop and establish procedures that will ensure compliance with federal, state, and local environmental and safety regulations. The asbestos consulting business is well established, and many qualified consultants are available to assist companies in preparing bid packages for abatement contractors and in preparing abatement procedures and removal specifications that comply with applicable regulations.

12

SUMMARY OF CLEAN WATER REGULATION

Statute:	33 USC § 1251 et seq.
Regulations:	40 CFR Parts 401 to 471 (Effluent Standards)
	33 CFR Parts 320 to 330 (Dredge and Fill)

The purpose of the Clean Water Act (CWA) is to "restore and maintain the chemical, physical, and biological integrity of the nation's waters" with the goal of attaining "fishable and swimmable" water conditions whenever possible. The Act requires states to develop water quality standards for each body of water, conduct assessments to determine whether standards are being met, identify sources of water pollution, and implement programs to control pollution.

Major federal clean water legislation includes the Federal Water Pollution Control Act of 1948, the Water Quality Act of 1965, the Federal Water Pollution Control Act Amendments of 1972, the Clean Water Act of 1977, and the Water Quality Act of 1987.

12.1 WATER DISCHARGE PERMITS

The keystone of the CWA is the National Pollutant Discharge Elimination System (NPDES) permit program, which regulates the discharge of pollutants (including storm water) into surface waters. Permit discharge limits are based on both available pollution control technology and the ability of the receiving stream to assimilate the discharged pollutants. Permit holders must monitor their discharges and report the results to the authorized environmental agency.

12.2 STORM WATER DISCHARGES

Section 402 of the Clean Water Act directs the Environmental Protection Agency (EPA) to establish a storm water discharge permit program to govern municipal and industrial storm water discharges into waters of the United States. The storm water permitting strategy developed by EPA is quite complex. Facilities subject to the storm water requirements may apply for an individual, group, or general permit, depending on a facility's specific circumstances.

EPA has decided to use general permits to address the vast majority of industrial storm water discharges.

12.3 OBLIGATIONS OF INDIRECT DISCHARGERS

An *indirect discharger* refers to a source introducing pollutants into a publicly owned treatment works (POTW). These dischargers are typically required to treat their effluent before it is discharged to the POTW, as provided by the terms of the "pretreatment permit" issued to the discharger by the POTW.

EPA is authorized to establish pretreatment standards applicable to certain indirect dischargers for controlling pollutants determined not to be susceptible to treatment by a POTW or that would interfere with the operation of the treatment works. Indirect dischargers subject to pretreatment standards may not discharge effluents into a POTW unless they comply with such standards.

12.4 WATER QUALITY STANDARDS

A water quality standard for a particular body of water consists of a designated use (such as public water supply, recreation, or agriculture) and criteria for various pollutants, expressed in numerical concentration limits necessary to support that use. Section 303 of the CWA requires every state to establish, and every three years to review, water quality standards for stream segments within the state.

Water quality standards may serve as the basis not only for imposing effluent limitations on point source dischargers, but also for establishing controls for nonpoint sources under water quality management plans.

12.5 SURFACE WATER RUNOFF

The largest source of water pollution in America's rivers, lakes, and streams is surface water runoff. This type of pollution is usually called ''nonpoint source'' pollution because it comes from a variety of sources rather than from a single source of discharge. Nonpoint source pollution is difficult to detect and control for that very reason. Indeed, many everyday activities and traditional land use practices contribute to nonpoint source pollution. Recognizing this problem, Congress amended the Clean Water Act in 1987 to require states to develop management programs to control nonpoint sources of pollution. States must identify categories of nonpoint sources that add significant pollution to waters and must develop a process for identifying ''best management practices'' and other measures to control these sources of pollution. The control of nonpoint source pollution is one of the major Clean Water Act challenges for the next century.

12.6 DREDGE AND FILL PERMITS

Historically, the United States Army Corps of Engineers has had authority, pursuant to the Rivers and Harbors Act of 1899, to regulate various activities impacting navigable waters. However, in recent years the authority provided by the Rivers and Harbors Act has been supplanted by the NPDES and Section 404 Permit Programs. Section 404 of the Clean Water Act requires a permit from the Corps of Engineers before dredge or fill materials can be discharged into waters of the United States, including wetlands. Federal responsibilities in regard to the program are divided in a complicated manner between the Corps and EPA. Although the Corps is the primary permitting agency, the EPA retains a major role in overseeing the permitting process. Several federal agencies, including the National Marine Fisheries Service and the United States Fish and Wildlife Service, are authorized to comment on dredge and fill permits and frequently comment on permits that may adversely impact endangered species of fish. The authorized state also must issue a water quality certification before a dredge and fill permit can be issued.

A major focal point of the dredge and fill permit program is the discharge of materials into ''wetlands.'' The inherent conflict between the need to protect our nation's wetlands and the continued pressure to undertake industrial development in wetland areas will result in a continued regulatory focus on dredge and fill activities.

12.7 GROUNDWATER PROTECTION AND CONTROL

There is no comprehensive federal regulatory program regarding the protection and control of groundwater. The only major federal program directed specifically to the protection of groundwater quality is the Underground Injection Control (UIC) program of the Safe Drinking Water Act. The Safe Drinking Water Act also requires states to develop programs for "well head protection."

A few states have enacted comprehensive groundwater protection programs, but most states have not yet taken any significant regulatory action. The protection of groundwater and drinking water supplies promises to be one of the major environmental issues of the next century, and further regulatory action on both the federal and state levels is expected.

12.8 SPILL PREVENTION AND NOTIFICATION

Section 311 of the CWA establishes notification requirements relating to the discharge or spilling of oil or hazardous substances into the waters of the United States. EPA has also promulgated regulations under the authority of this section of the Act, requiring facilities that store large quantities of oil products that could be discharged into navigable waters to prepare Spill Prevention, Control, and Countermeasure (SPCC) plans.

13

WATER QUALITY REGULATION

Water Quality Standard
Regulation: 40 CFR Part 131

Protection and enhancement of water quality will be one of the most significant environmental challenges of the next century. Water quality standards are the current mechanisms used by regulatory agencies to protect the quality of America's surface waters, and these will become more important in future years as the Environmental Protection Agency (EPA) moves away from technology-based effluent limitations and focuses on the restoration and protection of water quality in America's streams. This chapter summarizes some of the key concepts used by the states and the federal EPA to regulate water quality.

13.1 ESTABLISHMENT OF WATER QUALITY STANDARDS

A water quality standard for a particular body of water consists of designated beneficial uses and criteria for various pollutants required to protect those uses. The Clean Water Act (CWA) allows states to set their own water quality standards but requires that all beneficial uses and their protective criteria meet the goals of the Act.

State water quality standards are established through a three-step process. First, the state must designate the beneficial use for each body of water within

Table 13.1 Oregon water quality criteria for dissolved oxygen (DO)

Dissolved oxygen concentrations shall not be less than the following:

Type of Aquatic Habitat	DO Level (mg/l)
Salmonid Spawing Area (varies by location, gravel bed, dissolved oxygen, altitude, and temperature)	9.0–11.0
Cold Water Aquatic Life	8.0
Cool Water Aquatic Life	6.5
Warm Water Aquatic Life	5.5

the state. Fishing, recreation, and public water supply are typical beneficial uses. Once the designated uses are set, the state must then specify water quality criteria that will protect those designated uses. States typically incorporate into their regulations water quality criteria developed by EPA. Criteria can be numeric or narrative. Numeric criteria are statements of the acceptable numeric concentration of a specific pollutant required to support a given use. Narrative criteria are used to describe a desired water quality condition when there is not enough information on which to base a numeric criteria.

For example, Table 13.1 summarizes the Oregon water quality criteria for dissolved oxygen.

After a state develops its water quality protection program, it is sent to EPA for approval. If the state program is inadequate, EPA has authority under the CWA to promulgate standards for the state.

Under Section 401 of the Clean Water Act, an applicant for a federal permit for any activity that results in a discharge into the navigable waters of the United States must receive a water quality certification from the state. This ''401 certification'' is issued only after the state concludes that the proposed activity will comply with applicable water quality standards. A state may condition its certification in various ways, and those conditions are then incorporated into the federal license or permit.

13.2 ANTIDEGRADATION

Antidegradation refers to policies and procedures designed to prevent or minimize the deterioration of water quality below existing levels. EPA created the antidegradation policy in response to the CWA directive to "restore and maintain the chemical, physical, and biological integrity of the nation's waters." EPA's antidegradation policy requires states to develop and implement statewide antidegradation programs that meet the following objectives:

- Maintain existing in-stream uses and water quality levels.
- Maintain water quality that exceeds levels necessary to support fish, wildlife, and recreation, unless the state finds that lowering water quality is necessary to accommodate important economic and social development.
- Maintain and protect "outstanding national resource waters," which are generally considered to be water bodies that are important, unique, or ecologically sensitive (see 40 CFR § 131.12).

13.3 IDENTIFICATION OF WATER QUALITY STREAM SEGMENTS

Section 305(b) of the Clean Water Act requires each state to compile lists of water bodies in which the designated uses are being negatively impacted by excessive pollutant loadings and submit those lists in a water quality report to EPA every two years. Section 303(d) provides direction for the states in identifying those water bodies. In general, 303(d) requires states to determine which water bodies, by segment, fail to meet either numeric or narrative water quality standards because of inadequate pollution control, and then to list those water body segments as "water quality limited." It is common for states to list hundreds of water-quality-limited stream segments. States are required to use a combination of readily available scientific data and best professional judgment to determine which stream segments and lakes are to be listed. Examples of water quality standard violations include exceedances evidence of a numeric water quality standard, evidence of beneficial use impairment (e.g., decline in the population of a cold water fishery), evidence of a narrative standard violation, or a technical analysis (such as computer modeling) that suggests a decline in water quality.

Once the water-quality-limited stream segments are identified, states must then establish a priority ranking for these waters, taking into account the pollution severity and designated uses of the water. After these lists are finalized, states must develop total maximum daily loads (TMDLs) at whatever levels are necessary to achieve the applicable state water quality standards.

13.4 THE TMDL PROCESS

The TMDL is the total amount of particular pollutants or parameters such as temperature that sources can discharge into a receiving stream without violating water quality standards. The TMDL process is an important part of the water quality–based approach to pollution control designed by EPA. The process of calculating TMDL is essentially that of determining the capacity of each stream segment to assimilate pollutant discharges. The TMDL process establishes allowable loadings from the contributing point and nonpoint sources to a given water body and defines how much pollution reduction is necessary to achieve water quality standards. A simple formula for TMDLs is provided as follows:

$$\text{TMDL} = \text{nonpoint source pollution and background} + \text{point source waste load allocation} + \text{margin of safety}$$

The margin of safety accounts for the uncertainty of calculating pollutant loads and the estimated capacity of the receiving water body. Given the subjective nature of this concept, the margin of safety calculation is closely scrutinized by all parties interested in the TMDL process.

The TMDL process is designed to provide the following information:

- Inventory of all sources of a designated pollutant,
- An analysis of why pollution controls are ineffective,
- A plan to monitor and evaluate progress toward achieving the water quality standard,
- A list of pollution control strategies for reducing sources of the pollutant, and
- A prediction of the amount of time needed to restore and protect water quality.

The TMDL process is difficult, inexact, and controversial. A large number of states are currently defendants in litigation brought by environmental groups questioning the states' development of their 303(d) lists and subsequent TMDL evaluation and implementation actions. State environmental agencies are frustrated by the large number of stream segments on their 303(d) lists and the lack of reliable water quality information. The costs of TMDLs range greatly, depending on the size and complexity of the watershed, the number of point and nonpoint sources, and the degree of public participation. Establishing TMDLs can take from a few weeks to several years to complete.

Environmental managers should carefully monitor the 303(d) listings in their states to determine whether their facilities receiving stream segments are proposed for inclusion on a 303(d) list. Once a facility's stream segment is proposed for listing, the manager should take an active role in trying to influence the 303(d) and TMDL process, because this process can result in dramatic reductions in the allowable level of pollutant discharges from the facility.

13.5 CONTROL OF NONPOINT SOURCES

If water quality standards are ever to be met, nonpoint source pollution must be controlled. Nonpoint source discharges account for the major part of all water pollution. An estimated two-thirds of the nation's stream segments that fail to meet water quality criteria do so because of nonpoint discharges.

In 1987, Congress amended the CWA and added Section 319 to the Act, which directs states to develop management programs to control nonpoint sources of pollution. Section 319 requires states to prepare assessment reports to identify nonpoint source pollution problem areas and categories of nonpoint source pollution. Section 319 also directs states to identify Best Management Practices (BMPs) that can be used to control nonpoint source pollution and develop management programs to document how and when the states will address their nonpoint management concerns. Under the law, EPA has no authority to develop management programs for states that do not submit programs or whose programs are not approved by EPA. Consequently, regulation of nonpoint source pollution varies dramatically from state to state.

Table 13.2 BMP examples

AGRICULTURE	URBAN AREAS
• Animal waste management	• Flood storage
• Contour farming	• Runoff detention/retention
• Crop rotation	• Street cleaning
• Buffer strips	FORESTRY
• Livestock exclusion	• Ground cover maintenance
• Fertilizer management	• Limiting disturbed areas
CONSTRUCTION	• Log removal techniques
• Disturbed area limits	• Proper construction of haul roads
• Nonvegetative soil stabilization	• Removal of debris
• Runoff detention/retention	• Riparian zone management

Source: U.S. Environmental Protection Agency, 1991.

13.5.1 Best Management Practices (BMPs)

BMPs are methods and practices for preventing or reducing nonpoint source pollution to the level required to achieve water quality standards. BMPs are identified in Section 319 of the Clean Water Act as the primary mechanism to enable nonpoint sources to achieve water quality standards.

The EPA Water Quality Management Regulations more specifically define BMPs as:

> Methods, measures, or practices selected by an agency to meet its nonpoint source control needs. BMPs include but are not limited to structural and non-structural controls and operation and maintenance procedures. BMPs can be applied before, during, and after pollution-producing activities to reduce or eliminate the introduction of pollutants into receiving waters (40 CFR § 130.2(m)).

Several examples of Best Management Practices are outlined in Table 13.2.

13.5.2 Watershed Management

Recent thinking regarding nonpoint source pollution control focuses on the importance of using watershed management as a tool to meet the objectives of the Clean Water Act. Watershed management requires the involvement of the local ''stakeholders'' within a watershed—those individuals, interest groups, and agencies with an interest in properly managing and protecting watershed health by effectively controlling nonpoint source pollution. Coordinated watershed planning will allow these stakeholders to evaluate the needs, opportunities, and constraints within their watershed and agree upon practical and negotiated solutions to reduce nonpoint source pollution and protect water quality. Although the regulatory details of watershed management are still evolving, it is a concept that will be used often in the next century.

14

WATER DISCHARGES

Regulations:	40 CFR Parts 401 to 471	(Effluent Standards)
	33 CFR Parts 320 to 330	(Dredge and Fill)

The federal Clean Water Act (CWA) directs the Environment Protection Agency (EPA) and authorized states to administer several programs governing the discharge of wastewater into waters of the United States. Most industrial facilities have some type of wastewater discharge, and environmental managers should be familiar with the basic elements of water permitting. The water permitting programs that the environmental manager is most likely to encounter are summarized in this chapter.

14.1 NATIONAL POLLUTANT DISCHARGE ELIMINATION SYSTEM (NPDES) PERMITS

Section 402 of the CWA establishes the National Pollutant Discharge Elimination System (NPDES). The NPDES program requires permits for the discharge of pollutants from any point source into waters of the United States. The terms *pollutant, point source,* and *waters of the United States* are defined in detail in the federal regulations. In general, any material that is added to water (or, in some cases, any change in the characteristics of water, such as pH or temperature) constitutes a ''pollutant,'' and any discrete discharge point, such as a pipe, ditch, or other conveyance, is considered a ''point source.'' *Waters of the United States* is defined much more broadly than the traditional *navigable waters* and includes almost any surface water and the territorial seas, as well as adjacent wetlands.

NPDES permits are required for most activities that result in a discrete discharge of wastewater or other material to surface waters. Exceptions to the general NPDES permit requirement include the following activities:

- Discharges of pollutants into a publicly owned treatment works (POTW)
- Discharges of dredged or fill material regulated under Section 404 of the Clean Water Act
- Return flows from irrigated agriculture
- Discharges from nonpoint source agricultural and silvicultural activities

NPDES permits include specific effluent limitations for each pollutant discharged by a facility. The NPDES permit program is often administered by states whose programs meet certain conditions and are approved by EPA. Until a state establishes an approved NPDES program, EPA, through its regional offices, operates the NPDES permit program. NPDES permits are issued for fixed terms not to exceed five years. Whether issued by EPA or a state, all permits must meet certain minimum criteria. Permits must include provisions that require the periodic sampling and testing of effluents. The results of this testing are typically submitted to the authorized agency, on a monthly basis, on a standardized form called a Discharge Monitoring Report (DMR). A sample DMR form is provided in Figure 14.1.

NPDES permits also contain several pages of ''general conditions'' that apply to the permit. These general conditions, which often are ignored or taken lightly by permittees as nothing more than ''boilerplate'' or ''fine print,'' establish a broad range of legal obligations governing the operation and maintenance of the wastewater treatment system, monitoring, record keeping, and permit compliance. The environmental manager should take time to understand these conditions and incorporate those requirements into the facility's environmental management plan. A list of NPDES General Conditions is found in Appendix F.

14.1.1 Technology-Based Effluent Limitations

The key provisions of any NPDES permit are the effluent limitations, which specify how much pollution may be discharged by a facility. In developing technology-based effluent limitations or guidelines, EPA divided the universe of point source discharges into industrial categories and subcategories. EPA then established effluent limitations for specific pollutants in the various

PERMITTEE NAME/ADDRESS

Name: ______________

Address: ______________

____

Facility: ______________

Location: ______________

NATIONAL POLLUTANT DISCHARGE ELIMINATION SYSTEM *(NPDES)*

DISCHARGE MONITORING REPORT *(DMR)*

(2-18)	*(17-19)*
ABC-123	001
PERMIT NUMBER	DISCHARGE NUMBER

	MONITORING PERIOD						
	YEAR	MONTH	DAY		YEAR	MONTH	DAY
FROM	98	1	01	TO	98	2	01
	(20-21)	*(22-23)*	*(24-25)*		*(26-27)*	*(28-29)*	*(30-31)*

NOTE: Read instructions before completing this form

(32-37) PARAMETER		QUANTITY OR LOADING *(46-53)* AVERAGE	QUANTITY OR LOADING *(54-61)* MAXIMUM	UNITS	QUALITY OR CONCENTRATION *(38-45)* MINIMUM	QUALITY OR CONCENTRATION *(45-53)* AVERAGE	QUALITY OR CONCENTRATION *(54-61)* MAXIMUM	UNITS	NO. EX. *(62-63)*	FREQUENCY OF ANALYSES *(64-68)*	SAMPLE TYPE *(69-70)*
PRODUCTION	Sample Measurement			MDT/D						Monthly	--
	Permit Requirement									Monthly	
FLOW	Sample Measurement			MGAL/						Cont.	NA
	Permit Requirement									Cont.	NA
pH	Sample Measurement			--				--	0	Cont.	NA
	Permit Requirement							--		Cont.	NA
TEMPERATURE	Sample Measurement			°F						Cont.	NA
	Permit Requirement									Cont.	NA
TOTAL SUSPENDED SOLIDS (TSS)	Sample Measurement			LBS/D					0	7/7	24 HC
	Permit Requirement									7/7	24 HC
BIOCHEMICAL OXYGEN DEMAND (BOD)	Sample Measurement			LBS/D					0	7/7	24 HC
	Permit Requirement									7/7	24 HC
	Sample Measurement										
	Permit Requirement										

NAME/TITLE PRINCIPAL OFFICER:	I certify under penalty of law that I have personally examined and am familiar with the information submitted herein; and based on my Inquiry of those individuals immediately responsible for obtaining the information, I believe the submitted information is true, accurate and complete. I am aware that there are significant penalties for submitting false information, including the possibility of fine and imprisonment. See 18 U.S.C. 1001 and 33 U.S.C. 1319.	SIGNATURE OF PRINCIPLE EXECUTIVE OFFICER OR AUTHORIZED AGENT	TELEPHONE NUMBER:		
			DATE		
			98 Year	2 Month	13 Day

EPA Form 3320-1 (Rev.9-88)

Figure 14.1 National Pollutant Discharge Elimination System (NPDES) Discharge Monitoring Report (DMR)

categories and subcategories, based on the pollution control technologies available in the industry. In the absence of technology-based effluent limitations, permitting authorities may establish limitations through a case-by-case technology review, referred to as Best Engineering Judgment (BEJ).

The CWA provides that technology-based effluent limitations for existing sources should be based on the following three categories of technology review:

- Best Practicable Control Technology Currently Available (BPT). The BPT technology level is intended to represent the average of the best existing performance of facilities of various ages, sizes, and processes within the relevant point source category and applied to all point source discharges as of July 1, 1977.
- Best Conventional Pollution Control Technology (BCT) for certain conventional pollutants, such as biochemical oxygen demand, total suspended solids, fecal coliform, oil, and grease, and for pH. BCT limits were to be achieved by July 1, 1984.
- Best Available Technology Economically Achievable (BAT) for all toxic pollutants. EPA has published a list of 65 so-called ''priority pollutants'' which are or may be toxic. BAT limits were established for all ''nonconventional pollutants,'' which are those pollutants that have not been designated by EPA as either ''toxic'' or ''conventional.'' BAT limits must have been met by March 31, 1989.

EPA has also established effluent limitations for new sources, referred to as New Source Performance Standards (NSPS), based on the Best Available Demonstrated Control Technology. The rationale for this stringent technology standard for new sources is that new sources should be required to implement the best and most current wastewater technologies before they are allowed to operate. Effluent guidelines and standards for various industrial classifications are published at 40 CFR Parts 405 through 471. By way of example, selected sections of the effluent guidelines for the Organic Pesticide Chemicals Manufacturing Subcategory are included in Appendix G of this book.

14.1.2 Water Quality-Based Effluent Limitations

If technology-based effluent limitations are insufficient to meet water quality standards in a receiving stream, the CWA authorizes the imposition of more stringent ''water quality–based'' effluent limits. Under the law, water quality–based limits can be more restrictive than those based on technology. Translating water quality standards into specific permit limits requires, at a minimum, information about the flow and ambient quality of the receiving stream and the concentration of pollutants in the effluent. In the case of heavy metals or other pollutants whose effect on water quality is not complicated by biodegradation or other reactions over time, these limitations are usually set in a straightforward manner, calculated to ensure that the concentrations present in the discharge will not result in violations of water quality standards. For parameters such as biological oxygen demand (BOD) whose affect on water quality varies over time, the setting of water quality–based limitations is significantly more complicated, requiring the use of computer models or, alternatively, reliance on conservative assumptions that may result in unduly restrictive discharge limits. The difficulty of setting water quality–based limitations is further complicated where water quality in a stream segment is affected by more than one discharger and the burden of effluent reduction must be allocated among dischargers. Where there are multiple discharges to a single stream segment, the authorized agency must calculate the total maximum daily load (TMDL) for the segment and then perform a waste load allocation for each discharger. These procedures are discussed in more detail in Chapter 13 of this book.

Major revisions were made to the Clean Water Act in 1987 concerning situations where state water quality standards and/or technology-based effluent limitations have not reduced toxic pollutant concentrations to acceptable levels. First, states must identify waters that, after the application of technology-based effluent limitations and categorical pretreatment standards, cannot attain or maintain either state water quality standards or a level of water quality that protects human health and the environment. Next, individual control strategies (ICSs) must be developed for each identified point source that is contributing to the failure to meet the water quality standard. Effluent limitations are then established, based on the individual control strategies. The ICS must ensure that the applicable water quality standard will be achieved within three years of the time the ICS is was established.

14.1.3 NPDES Permitting Practices

There are several practical matters for environmental managers to consider when negotiating the provisions of an NPDES permit. First, because most effluent limitation guidelines specify an allowable mass discharge for each unit of production, it is important to establish the maximum production rates of the facility. Production is typically defined as ''off-the-machine production,'' and the calculation of this production should include product rejects and recycled material in addition to the actual amount of product manufactured for sale by the facility.

It also is important for the facility to correctly establish at what point its effluent discharge will be measured to determine compliance with the facility's effluent limitations. In order to properly establish this point of compliance, the environmental manager must understand the overall design of the facility and where the different waste streams (e.g., process wastewater, cooling water, and storm water) are discharged. Permit compliance testing should be done at a point that fairly represents the entire wastewater discharge of the facility.

The manner in which a permittee must monitor for compliance with permit effluent limitations is almost as important as the limitations themselves. Specific monitoring requirements are established by the permit writer when the permit is issued, and the regulatory agencies have a great deal of discretion in establishing what monitoring methods and frequency of monitoring will be required by the permit. Environmental managers should be very sensitive to the high costs of analytical testing. The economic impact of frequent monitoring can be a powerful argument against the imposition of daily, weekly, or even monthly monitoring requirements for many pollutants. For example, a requirement to test a toxic pollutant such as dioxin on a monthly, rather than a quarterly, basis can translate into many thousands of dollars in additional laboratory costs over the five-year life of an NPDES permit.

Finally, in regard to NPDES permits issued by EPA, the agency has developed specific procedures for complying with the federal Endangered Species Act (ESA). EPA will submit all proposed permits to the U.S. Fish and Wildlife Service (FWS) and the National Marine Fisheries Service (NMFS) to determine whether the proposed permit threatens any endangered species. If the federal agencies agree that the proposed permit does not threaten endangered species, the permit may be issued. If the proposed permit does threaten endangered species, the permittee then becomes subject to the time-consuming and confused regulations established by the ESA to determine whether there is any way to issue the permit and still protect the endangered species. The increase in the number of listed species in America's waterways makes it more likely that the ESA will become a major factor in the issuance of federal NPDES permits. Many states are already considering ESA impacts in identifying biologic parameters for inclusion in NPDES permits. The selection of vertebrate and invertebrate species for lethal dose testing can be subjective. This can result in permit compliance conditions that are difficult, if not impossible to meet.

14.2 INDUSTRIAL DISCHARGES INTO PUBLICLY OWNED TREATMENT WORKS

Industrial facilities often discharge their wastewater into public sewer systems, which are commonly referred to as publicly owned treatment works

(POTW). POTWs are usually operated by cities or local sewerage agencies. Dischargers to POTW are typically required to treat their effluent prior to discharge, as provided by the terms of the ''pretreatment permit'' issued to the discharger by the POTW.

POTWs are governed by the General Pretreatment Regulations adopted by EPA. These regulations generally require POTWs to develop and enforce a pretreatment program that contains the following elements:

- A POTW must enforce narrative standards, numerical criteria, and categorical standards that EPA has developed to regulate indirect discharges. The POTW may also develop and enforce its own discharge limitations, commonly referred to as ''local limits.''
- The POTW must issue individual discharge permits to all ''Significant Industrial Users'' in a timely manner. Most industrial facilities that discharge to POTW will be considered Significant Industrial Users and, therefore, require a pretreatment permit.
- The POTW must develop a program to sample and analyze the effluents discharged by industrial users.
- The POTW must review reports submitted by industrial users and identify all permit violations.
- The POTW must investigate all instances of industrial user noncompliance and take appropriate enforcement action.
- The POTW must develop and maintain a data management system designed to track the status of industrial users, discharge characteristics, and compliance. The POTW must also publish an annual report describing the pretreatment program.
- Industrial user permits usually contain effluent limitations, monitoring and reporting requirements, and special and general conditions.
- Violators of industrial user permits can be assessed civil and/or criminal penalties.

Cities and local sewerage agencies often are lax in their compliance with the requirements of the pretreatment regulations. In such cases, the burden falls on the industrial user to work with the POTW to ensure that the user has obtained all necessary regulatory approvals.

14.3 STORM WATER DISCHARGES

Section 402 of the Clean Water Act directs EPA to establish a storm water discharge permit program to govern municipal and industrial storm water discharges. Many industrial facilities discharge their storm water under the authority of their NPDES permits. For those facilities not already governed by permits, the EPA has promulgated regulations for issuing storm water permits for any separate storm water discharge associated with industrial activity. The regulations provide a very complicated definition of "industrial activity." In general, a storm water permit is required if the facility is included in a broad list of categories and Standard Industrial Classification (SIC) codes identified by EPA, and storm water from rain or snowmelt leaves

Table 14.1 SIC codes designating facilities subject to storm water regulation

SIC Code	Description
10	Metal Mining
12	Coal Mining
13	Oil and Gas Extraction
14	Nonmetallic Minerals, Except Fuels
20	Food and Kindred Products
21	Tobacco Products
22	Textile Mill Products
23	Apparel and Other Textile Products
24	Lumber and Wood Products
25	Furniture and Fixtures
26	Paper and Allied Products
27	Printing and Publishing
28	Chemical and Allied Products
29	Petroleum and Coal Products
30	Rubber and Miscellaneous Plastics Products
31	Leather and Leather Products
32	Stone, Clay, and Glass Products
33	Primary Metal Industries
34	Fabricated Metal Products
35	Industrial Machinery and Equipment
36	Electronic and Other Electric Equipment
37	Transportation Equipment
38	Instruments and Related Products
39	Miscellaneous Manufacturing Industries
40	Railroad Transportation
41	Local and Interurban Passenger Transit
42	Trucking and Warehousing
43	U.S. Postal Service
44	Water Transportation
45	Transportation by Air
5015	Motor Vehicle Parts, Used
5093	Scrap and Waste Materials
5171	Petroleum Bulk Stations and Terminals

the site through a point source and reaches surface water. The environmental manager at a facility that falls within the SIC codes listed in Table 14.1 should determine whether the facility is subject to the storm water permit program.

The storm water permitting strategy developed by EPA is quite complex. Facilities subject to the storm water requirements may apply for an individual, group, or general permit, depending on a facility's specific circumstances. EPA's Notice of Intent form for storm water discharges is provided in Figure 14.2.

14.3.1 Individual Permits

Any dischargers subject to the storm water regulations may apply for an individual permit. However, only large industrial facilities that discharge significant amounts of storm water will likely receive individual permits.

14.3.2 Group Permits

EPA regulations provide that a group of at least four dischargers who belong to the same effluent guideline subcategory, or whose activities are sufficiently similar, may apply for a group permit. In practice, very few storm water dischargers have applied for group permits, owing to the wide range of general permits promulgated by EPA and state regulatory gencies.

14.3.3 General Permits

As a result of the large number of regulated storm water discharges and the administrative burden associated with processing and issuing thousands of individual or group permits, EPA and the authorized states have issued a series of general permits to address particular industrial activities. Most of the general permits contain the following requirements:

- A Storm Water Pollution Control Plan must be developed by the permittee. The plan must include a detailed site description, site controls and record keeping, and internal reporting procedures. Permittees also are required to implement the following storm water best management practices:
- Containment of hazardous materials to prevent leaks and spills
- Methods to minimize oil and grease contamination, such as the use of oil/water separators, booms, and skimmers
- Proper procedures for handling and disposing of waste chemicals

THIS FORM REPLACES PREVIOUS FORM 3510-6 (8-92)
See Reverse for Instructions

Form Approved. OMB No. 2040-0086
Approval expires: 8-31-98

NPDES FORM

EPA

United States Environmental Protection Agency
Washington, DC 20460

Notice of Intent (NOI) for Storm Water Discharges Associated with Industrial Activity Under a NPDES General Permit

Submission of this Notice of Intent constitutes notice that the party identified in Section II of this form intends to be authorized by a NPDES permit issued for storm water discharges associated with industrial activity in the State identified in Section III of this form. Becoming a permittee obligates such discharger to comply with the terms and conditions of the permit. ALL NECESSARY INFORMATION MUST BE PROVIDED ON THIS FORM.

I. Permit Selection: You **must** indicate the NPDES Storm Water general permit under which you are applying for coverage. Check one of these.

Baseline Industrial ☐ Baseline Construction ☐ Multi-Sector (Group Permit) ☐

II. Facility Operator Information

Name: ______ Phone: ______

Address: ______ Status of Owner/Operator: ☐

City: ______ State: ______ ZIP Code: ______

III. Facility/Site Location Information

Name: ______

Is the facility located on Indian Lands? (Y or N) ☐

Address: ______

City: ______ State: ______ ZIP Code: ______

Latitude: ______ Longitude: ______ Quarter: ______ Section: ______ Township: ______ Range: ______

IV. Site Activity Information

MS4 Operator Name: ______

Receiving Water Body: ______

If you are filing as a co-permittee, enter storm water general permit number: ______

SIC or Designated Activity Code: Primary: ______ 2nd: ______

Is the facility required to submit monitoring data? (1, 2, 3, or 4) ☐

If You Have Another Existing NPDES Permit, Enter Permit Number: ______

Multi-Sector Permit Applicants Only:

Based on the instructions provided in Addendum H of the Multi-Sector permit, are species identified in Addendum H in proximity to the storm water discharges to be covered under this permit, or the areas of BMP construction to control those storm water discharges? (Y or N) ☐

Will construction (land disturbing activities) be conducted for storm water controls? (Y or N) ☐

Is applicant subject to and in compliance with a written historic preservation agreement? (Y or N) ☐

V. Additional Information Required for Construction Activities Only

Project Start Date: ______ Completion Date: ______ Estimated Area to be Disturbed (in Acres): ______

Is the Storm Water Pollution Prevention Plan in compliance with State and/or Local sediment and erosion plans? (Y or N) ☐

VI. Certification: The certification statement in Box 1 applies to all applicants.
The certification statement in Box 2 applies only to facilities applying for the Multi-Sector storm water general permit.

BOX 1

ALL APPLICANTS:

I certify under penalty of law that this document and all attachments were prepared under my direction or supervision in accordance with a system designed to assure that qualified personnel properly gather and evaluate the information submitted. Based on my inquiry of the person or persons who manage the system, or those persons directly responsible for gathering the information, the information submitted is, to the best of my knowledge and belief, true, accurate, and complete. I am aware that there are significant penalties for submitting false information, including the possibility of fine and imprisonment for knowing violations.

BOX 2

MULTI-SECTOR STORM WATER GENERAL PERMIT APPLICANTS ONLY:

I certify under penalty of law that I have read and understand the Part I.B. eligibility requirements for coverage under the Multi-Sector storm water general permit, including those requirements relating to the protection of species identified in Addendum H.

To the best of my knowledge, the discharges covered under this permit, and construction of BMPs to control storm water run-off, are not likely to and will not likely adversely affect any species identified in Addendum H of the Multi-Sector storm water general permit or are otherwise eligible for coverage due to previous authorization under the Endangered Species Act.

To the best of my knowledge, I further certify that such discharges, and construction of BMPs to control storm water run-off, do not have an effect on properties listed or eligible for listing on the National Register of Historic Places under the National Historic Preservation Act, or are otherwise eligible for coverage due to a previous agreement under the National Historic Preservation Act.

I understand that continued coverage under the Multi-Sector general permit is contingent upon maintaining eligibility as provided for in Part I.B.

Print Name: ______ Date: ______

Signature: ______

EPA Form 3510-6 (8-98)

Figure 14.2 Notice of Intent for Storm Water Discharges Associated with Industrial Activity Under an NPDES General Permit form

- Erosion and sediment control procedures
- Debris control
- Storm water diversion to prevent exposure to pollutants
- Covering of chemical treatment storage and disposal areas to prevent storm water contamination
- Housekeeping practices to prevent exposure of pollutants to storm water
- Spill prevention and response procedures
- Preventive maintenance
- Employee orientation and education program
- Storm water must be monitored by taking grab samples twice a year for contaminants specified in the permit.
- Storm water discharge limits established by the permit must be met within a year of permit issuance.

General permits are also issued for construction activities, including cleaning, grading, and excavation activities, that disturb five or more acres of land. These permits typically require the contractor to prepare and implement an approved erosion control plan prior to beginning construction activities.

14.4 DREDGE AND FILL PERMITS

Section 404 of the Clean Water Act authorizes the United States Army Corps of Engineers to issue permits before dredged or fill materials can be discharged into waters of the United States, including wetlands. Indeed, much of the confusion and complexity regarding dredge and fill permitting relates to the regulation of discharges into jurisdictional wetlands. The Corps of Engineers regulations broadly define *dredged material* as ''material that is excavated or dredged from waters of the United States.'' *Fill material* is ''any material used for the primary purpose of replacing an aquatic area with dry land or changing the bottom elevation of a waterbody'' (33 CFR § 323.2).

The existing dredge and fill permit program divides authority between the Corps of Engineers and EPA. Although the Corps is the primary permitting agency, EPA retains a major role in overseeing the permitting process. Several other federal agencies are authorized to comment on permit applications. In particular, the U.S. Fish and Wildlife Service and the National Marine Fisheries Service are active in working with the Corps to ensure that dredge and fill activities do not threaten any listed endangered species.

The Clean Water Act also authorizes the Corps of Engineers to delegate a portion of its permitting authority to the states, although it is rare for a state to receive fully delegated 404 permitting authority. State program requirements for dredge and fill permitting programs have been published by EPA at 33 CFR Part 323. It is more common for states to develop their own dredge and fill permitting programs, in which case facilities must meet both the federal and state requirements.

Once a determination is made that an activity results in a discharge of dredged or fill material into the waters of the United States, the Act provides the following three alternatives:

- No permit will be required, because the activity falls within an exemption,
- No individual permit will be required, because the activity falls within the terms of a Section 404 General Permit, or
- An individual Section 404 Permit will be required.

These alternatives are briefly summarized in the following paragraphs.

14.4.1 Exempted Activity

Section 404(f) of the Clean Water Act exempts a variety of activities from the dredge and fill permit requirements:

- Normal farming, silviculture, and ranching activities
- Maintenance, including emergency reconstruction, of dikes, dams, levies, riprap, breakwaters, causeways, and bridge abutments
- Construction or maintenance of farm or stock ponds or irrigation ditches and the maintenance of drainage ditches
- Construction of temporary sedimentation basins on a construction site
- Construction or maintenance of farm roads or forest roads, so long as such roads are constructed and maintained in accordance with best management practices
- Activities conducted pursuant to a state water quality management plan

14.4.2 General Permits

Dredge or fill activity may be authorized by "general" permits. General permits are blanket authorizations for specified fill activities that will cause only minimal adverse impact to the environment. General permits are issued on a nationwide or regional basis. By regulation, the Corps of Engineers has established 39 different types of nationwide permits. Nationwide permits are issued for a five-year period and are available only in states that have certified that the activities authorized will comply with state water quality requirements.

One of the most widely used and controversial nationwide permits is Permit 26, which fully exempts from the permit requirement any activities that modify "isolated" wetland areas of less than one-third of an acre. Notice to the Corps is required for activities affecting isolated wetlands larger than one-third of an acre but smaller than three acres. In general, isolated wetlands are those not connected with a surface water system. The Corps is proposing to phase out Nationwide Permit 26 and replace it with a series of activity-based permits.

14.4.3 Individual Permits

If a dredge or fill activity is neither an exempt activity nor authorized by a nationwide or regional permit, an individual 404 Permit must be issued by the Corps of Engineers before any dredge or fill activities can proceed. Obtaining an individual 404 Permit can be a time-consuming and costly process, and the environmental manager should account for these delays in the project permitting schedule. An application for a 404 Permit must include, among other things:

- A detailed description of the proposed project, including maps and drawings,
- Location and purpose of the project,
- Schedule for construction, and
- List of authorizations required by either federal, state, or local agencies.

The application must also specify the source of the dredged or fill material to be discharged, the purpose of the discharge, a description of the type, composition, and quantity of the material, the method of transportation and disposal of the material, and the location of the disposal site.

Before a 404 Permit is issued, the Corps of Engineers must ensure that the proposed discharge is protective of both the aquatic ecosystem and the human use of the area. The agency also conducts a wide-ranging "public interest review," during which the "benefits which reasonably may be expected to accrue from the proposal must be balanced against the reasonably foreseeable detriments" (33 CFR § 320.4(a)(1)).

The dredge and fill permit program provides that impacts to wetlands must be avoided to the maximum extent possible. The Corps of Engineers must also seek to minimize the project's impacts on wetlands through project modifications and protective permit conditions. Mitigation is required for all unavoidable impacts to wetlands and is the final step of the dredge and fill permitting process. If there are no practicable alternatives to the proposed dredge and fill activity and after minimization there are still unavoidable adverse impacts, the applicant must provide compensatory mitigation. Compensatory mitigation, such as constructing replacement wetlands or enhancing deteriorated wetlands, should be undertaken in areas adjacent or contiguous to the filled wetlands. If no on-site mitigation is available, off-site mitigation can be undertaken in the same geographic area, preferably within the same watershed. Replacement wetlands are acceptable only if they achieve a minimum of a one-to-one functional replacement. Wetlands that are restored, created, enhanced, or preserved expressly for the purpose of providing compensatory mitigation for wetlands projects, may be credited in a ''mitigation bank'' and later ''withdrawn'' to satisfy mitigation requirements imposed on permitted dredge and fill activities.

As a result of these provisions and the potential involvement of several federal agencies, the administration of the Section 404 permitting process is very subjective. Industrial facilities should plan their activities in such a way as to avoid dredge and fill activities in wetlands or other waters of the United States.

15

THE SAFE DRINKING WATER ACT

Statute:	42 USC §§ 300f to 300j–26
Regulations:	40 CFR Parts 141 to 148

The Safe Drinking Water Act (SDWA) was initially enacted by Congress in 1974. Major amendments were made in 1986, and the Act was reauthorized and amended again in 1996. The 1996 SDWA Amendments contain a significant number of new regulatory requirements for the Environmental Protection Agency (EPA), the states, water suppliers, and the public. EPA is in the process of promulgating all of the regulations required by the Amendments.

The SDWA is subdivided into six parts. This summary is intended to provide only a general overview of the Act. Although the Act grants primary enforcement responsibility for the program to the states, this chapter does not discuss applicable state programs.

15.1 APPLICABILITY OF THE SDWA

Part A of the Act contains the definitions that outline the applicability of the law. According to its terms, the requirements of the SDWA apply to public water systems. The Act defines a "public water system" as a system providing piped water to the public for "human consumption" if such system has at least "15 service connections or regularly serves at least 25 individuals.'' Although neither the statute nor the regulations define the terms *drinking water* or *human consumption*, at least one court decision has concluded that water for human consumption includes not only water that is used for drinking, but also water that is used for bathing and showering, cooking and dish-washing, and maintaining oral hygiene. Any industrial facility that provides water for human consumption to 25 or more employees or other individuals is operating a public water system and is subject to the provisions of the SDWA. Environmental managers for these facilities should ensure that they comply with the detailed requirements of the Act, because failure to comply can subject the facility to both regulatory penalties and the possibility of civil liability relating to harm suffered by employees who consume impure water. Although many industrial facilities receive their drinking water from municipal water systems, a surprisingly large number of facilities, especially in remote locations, supply their employees with drinking water.

15.2 PUBLIC WATER SYSTEMS

The SDWA establishes a detailed regulatory program governing public water systems. The requirements of the regulations are summarized in the following paragraphs.

Table 15.1 Maximum contaminant levels for inoganic contaminants[a]

Contaminant	MCL (mg/l)
Fluoride	4
Asbestos	7 million fibers/liter (longer than 10 μm)
Barium	2
Cadmium	0.005
Chromium	0.1
Mercury	0.002
Nitrate	10 (as nitrogen)
Nitrite	1 (as nitrogen)
Total Nitrate and Nitrite	10 (as nitrogen)
Selenium	0.05
Antimony	0.006
Beryllium	0.004
Cyanide (as free cyanide)	0.2
[Reserved]	
Thallium	0.002

[a] See 40 CFR § 141.62.

Primary Drinking Water Standards The SDWA establishes National Primary Drinking Water Regulations, which apply to every public water system. Maximum contaminant levels (MCLs) are established for a variety of organic and inorganic chemicals, turbidity, coliform bacteria, and various measures of radioactivity. By way of example, the MCLs for inorganic contaminants are listed in Table 15.1.

These regulations also specify monitoring and analytical requirements, reporting, public notification and record-keeping obligations, and special rules relating to filtration, disinfection, and the control of lead and copper. In regard to future regulation of drinking water contaminants, EPA will decide which contaminants to regulate, based on data regarding the adverse health effects of a contaminant, its occurrence in public water systems, and the projected reduction in risk through regulation.

Secondary Drinking Water Standards National Secondary Drinking Water Regulation's (NSDWR) have also been promulgated by EPA to control drinking water contaminants that primarily affect the aesthetic qualities of drinking water. The NSDWR are not federally enforceable but, rather, are intended as guidelines for the states. The secondary maximum contaminant containment levels for public water systems are summarized in Table 15.2.

Monitoring Requirements The federal SDWA regulations establish detailed monitoring and analytical requirements relating to the sampling of various contaminants required by the law.

Table 15.2 Secondary maximum contaminant levels[a]

Contaminant	Level
Aluminum	0.05 to 0.2 mg/l
Chloride	250 mg/l
Color	15 color units
Copper	1.0 mg/l
Corrosivity	Noncorrosive
Fluroide	2.0 mg/l
Foaming agents	0.5 mg/l
Iron	0.3 mg/l
Manganese	0.05 mg/l
Odor	3 threshold odor number
pH	6.5–8.5
Silver	0.1 mg/l
Sulfate	250 mg/l
Total dissolved solids (TDS)	500 mg/l
Zinc	5 mg/l

[a] See 40 CFR § 143.3

Operator Certification The 1996 SDWA Amendments direct EPA to publish guidelines for the states, that establish minimum standards for certifying operators of public water systems.

Prohibition on Use of Lead Pipes, Solder and Flux, and Water Coolers Containing Lead The SDWA prohibits the use of lead pipes, solder, or flux in the installation or repair of any public water system or any plumbing in a residential or nonresidential facility providing water for human consumption that is connected to a public water system. The Act also prohibits the manufacture and sale of water coolers containing lead.

Cross Connection Control Requirements For purposes of SDWA compliance, a "cross connection" is defined as any link or channel between the piping that carries drinking water and the piping that carries process water or other substances. State drinking water programs typically require water suppliers to undertake programs for controlling and eliminating cross connections. Cross connections are typically controlled through the installation of approved backflow devices that are tested and reinspected annually to ensure their integrity. From a practical standpoint, the environmental manager's greatest challenge in regard to cross connections is to prevent future cross connections by training maintenance, construction, and operating personnel as to the importance of not creating additional cross connections within the facility. In addition, potable water lines should be clearly labeled and color coded, if feasible, to help prevent the possibility of inadvertent cross connections.

General Reporting, Public Notification, and Record Keeping EPA regulations establish requirements for reporting, public notification, and recordkeeping. These regulations generally require suppliers of water to report test results to the state, notify users of any failure to comply with an applicable maximum contaminant level, and keep records of monitoring results and other relevant information. Beginning in 1999, water systems will be required to prepare and distribute annual reports to their customers. These reports will contain information regarding the source of the drinking water and the quality of the source, information relating to any detected contaminants in the drinking water, and a plain-language explanation of the health effects of any contaminants. The following is an example of EPA's mandatory health-effects language for the chemical benzene.

> Benzene. The United States Environmental Protection Agency (EPA) sets drinking water standards and has determined that benzene is a health concern at certain levels of exposure. This chemical is used as a solvent and degreaser

of metals. It is also a major component of gasoline. Drinking water contamination generally results from leaking underground gasoline and petroleum tanks or improper waste disposal. This chemical has been associated with significantly increased risks of leukemia among certain industrial workers who were exposed to relatively large amounts of this chemical during their working careers. This chemical has also been shown to cause cancer in laboratory animals when the animals are exposed at high levels over their lifetimes. Chemicals that cause increased risk of cancer among exposed industrial workers and in laboratory animals also may increase the risk of cancer in humans who are exposed at lower levels over long periods of time. EPA has set the enforceable drinking water standard for benzene at 0.005 parts per million (ppm) to reduce the risk of cancer or other adverse health effects which have been observed in humans and laboratory animals. Drinking water which meets this standard is associated with little to none of this risk and should be considered safe (40 CFR § 141.32(e)(5)).

15.3 ENFORCEMENT RESPONSIBILITIES

The environmental manager should be aware that the state has the primary enforcement responsibility for public water systems if the state adopts a contamination control program that conforms to federal requirements. Currently, 49 states are primarily responsible for administering the public drinking water protection program. In order for a state program to conform with federal requirements, the state must adopt primary drinking water regulations that are at least as stringent as the National Primary Drinking Water Regulations promulgated by EPA. Conformity with federal requirements also requires the state to adopt and implement adequate procedures for the enforcement of state regulations.

When a state fails to ensure the adequate enforcement of drinking water

Table 15.3 Categories of UIC wells[a]

Class 1	All wells that dispose of solid wastes beneath the lowermost formation where there is an underground drinking water source within a quarter mile radius of the well
Class II	Injection of fluids associated with oil and natural gas production activities
Class III	Injection of fluids associated with certain mineral extraction activities
Class IV	All wells that dispose of hazardous or radioactive wastes into or above a formation containing an underground source of drinking water within one-quarter mile of the well
Class V	All injection wells not included in Classes I, II, III, and IV

[a] See 40 CFR Parts 144 to 148.

regulations, EPA may commence a court action to require compliance on the part of the state. Although states are permitted to grant variances and exemptions from the regulations, EPA may intervene directly if a state abuses its discretionary power in granting such variances and exemptions.

15.4 PROTECTION OF UNDERGROUND SOURCES OF DRINKING WATER

Part C of the Safe Drinking Water Act is designed to protect underground sources of drinking water. Generally speaking, the regulations define "underground injection" as the subsurface injection of fluids through a bored, drilled, or driven well. States have been granted primary authority to regulate underground injections. EPA has issued regulations governing underground injection activities, which act as minimal requirements for relevant state plans. The regulations classify wells in five categories and provide different requirements for each class of well. These Underground Injection Control (UIC) well categories are summarized in Table 15.3.

15.5 EMERGENCY POWERS

Part D of the SDWA gives the EPA emergency powers in the event that contamination of a public water system is likely to create an "imminent and substantial endangerment" to the health and safety of persons. These powers allow EPA to issue orders necessary to protect human health and commence court actions for appropriate relief, including restraining orders or temporary injunctions.

Supplying drinking water to employees is a highly regulated activity, and any facility that assumes this responsibility must pay close attention to the requirements of the SDWA.

16

GROUNDWATER

Environmental management issues relating to groundwater have increased tremendously in recent years. The rising concern over contamination of this precious resource has led to increasing regulation by both state and federal agencies. Industry itself has recognized the potentially huge liabilities associated with groundwater contamination and has taken steps to reduce the risk of contamination associated with the design and operation of its facilities.

The environmental manager may become involved with groundwater assessments in a variety of different ways. Although groundwater studies are most often associated with determining the extent and location of existing groundwater contamination, they also are used to determine how an industrial operation can be designed and constructed to protect groundwater quality. The study of groundwater and hydrology is a complex science, and you will likely require supplemental expert assistance when evaluating any groundwater issue.

16.1 GROUNDWATER BASICS

The earth's surface is composed of various soils and rock formations. Some formations hold water because they are permeable or porous. When water is present in sufficient quantities, it can be pumped to the surface for uses above ground. This underground supply of water comes from a geologic formation called an aquifer. An aquifer is a formation that contains water and has the potential to supply water to a well or spring at the surface. It is important to note that layers of water-yielding and non–water–yielding formations may occur, thus resulting in aquifers at different elevations, separated by non-porous rock formations. Groundwater travels through an aquifer in a direction determined by the hydraulic gradient. That is, pressure is higher on one side of the formation than the other. The hydraulic gradient may be due to elevation changes (downhill flows), or it may be due to fluctuations in a surface water level, such as a lake or river fluctuating with seasonal variances in level and pushing water through adjacent ground at different flow rates. The concept is illustrated in Figure 16.1.

The velocity of the groundwater flow is dependent on the magnitude of the groundwater hydraulic gradient and the permeability of the soil formation. Permeability is a measure of the friction imposed on the water flow, expressed in distance traveled per unit of time. Permeability varies greatly. For example, clay or silt may yield a permeability of 0.1 to 10 feet/year, whereas sand or gravel may have permeabilities as great as 100,000 feet per year or greater. In general, groundwater moves very slowly, especially in relation to surface water flow.

The quality of groundwater also varies tremendously. Water is a very good solvent and many elements or substances dissolve naturally in water. Groundwater typically dissolves minerals from the surrounding soil and rock formations, which accounts for the natural presence of contaminants such as iron or silver in groundwater. The addition of contaminants can affect the

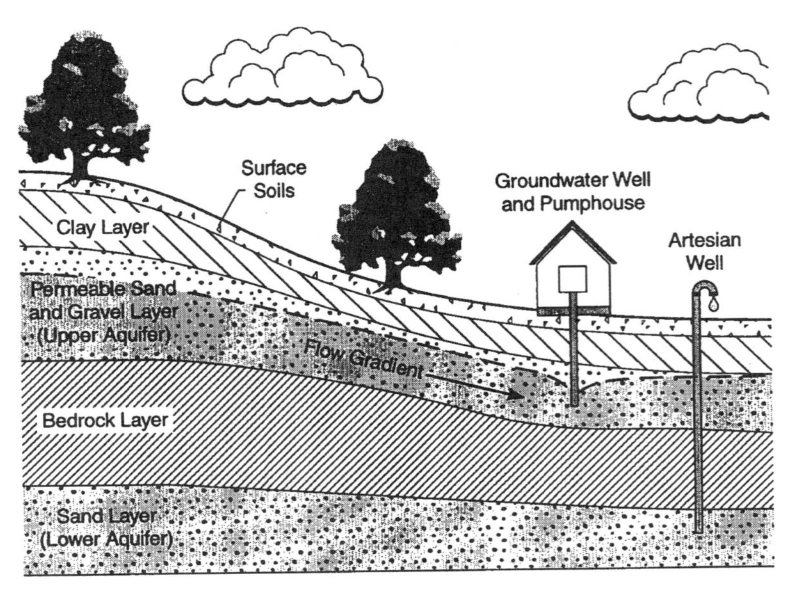

Figure 16.1 Groundwater aquifer and well

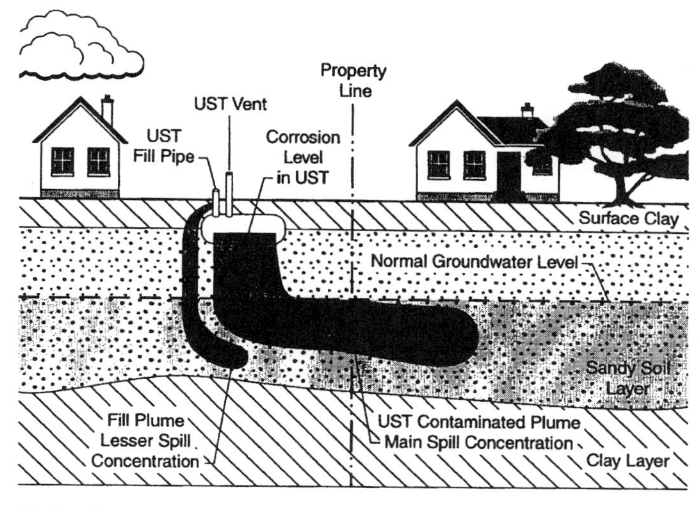

Figure 16.2 Groundwater contamination plume from underground tank leaks

pH of groundwater, which will impact the ability of the groundwater to dissolve other substances.

Groundwater flow and the ability of the water to transport chemicals are just two of the most important aspects of groundwater hydrology. A chemical spill that reaches groundwater will travel and spread. This "plume" of contamination may reach the groundwater of neighboring landowners. This concept is illustrated in Figure 16.2. The definition and remediation of these plumes is the focal point of most Superfund cleanup efforts. Groundwater remediation is not always possible, because of the complexities of groundwater flow and plume location. When remediation is feasible, the process is often slow, costly, and fraught with uncertainty.

16.2 STUDIES AND PERMITS

There are several environmental management issues that may require some level of groundwater study. Groundwater studies can be required before a facility can obtain a permit to use groundwater for process or drinking water. Groundwater monitoring is also used to establish the baseline condition of groundwater or to determine what impacts an existing surface lagoon or other process facility has on groundwater quality. Groundwater studies typically gather scientific information regarding flow, depth of aquifers, water quality, and seasonal impacts. The appropriate local or state agencies are contacted for information regarding possible contamination from neighboring sources and other relevant groundwater history. An assessment of aerial photos, surrounding property records, recorded spill events, evidence of leaking tanks, and other relevant information can also be helpful in determining the appropriate level of groundwater study.

If monitoring or test wells are to be established, the location of such wells is critical. An experienced hydrologist familiar with the site should be involved in designing the project and identifying well locations. Pump tests of wells are used to determine quantity production rates, flow direction, and water quality. On a large site, these characteristics may vary from one area to another because of geologic formations or surface water impacts. If you are conducting testing for review by an environmental agency, the agency should be involved in the review and approval of your initial plans.

16.3 USE AND TREATMENT

A facility that plans to use groundwater for process water, cooling water, or drinking water must first obtain a water rights certificate or permit. Water rights are discussed in detail in Chapter 17 of this book. Depending on the intended use of the groundwater, the facility may also have to pretreat or condition the water prior to use. Drinking water uses are subject to detailed standards, as described in Chapter 15. If the water is to be used for process or production purposes, the environmental manager should review the water quality requirements with the appropriate operations and technical staff at the facility. For example, if the water is to be used as boiler feed water, silica and hardness levels may be of concern. Although these substances can be treated with demineralizers, that can be a costly process. Another example is the use of groundwater in a washing process, in which case the turbidity of the water can be an issue. Turbidity can be treated by sand filters or other filtration methods, but remember, such treatment typically results in a byproduct waste stream that may need a water discharge permit. The treatment of contaminated water is discussed later in this chapter.

16.4 GROUNDWATER PROTECTION MONITORING

Potential groundwater contamination sources such as landfills, underground tanks, and treatment lagoons are required by regulation to meet stringent requirements to protect groundwater. Even in the absence of regulatory requirements, environmental best management practices dictate that all processes or chemical storage units with potential to impact groundwater should be designed and operated in a manner to prevent possible groundwater contamination. Modern techniques for designing landfills and lagoons include combinations of man-made liner fabrics and compacted clays with low permeability to protect groundwater. Collection systems for leachate, the liquid that drains through a landfill, also prevent the liquid from collecting above the liner, further reducing the risk of groundwater contamination.

New underground tanks are required to be designed with double walls and with leak detection systems. Existing units must be upgraded with leak detection based on the regulatory requirements outlined in Chapter 18. However, facilities with underground storage tanks (USTs) installed prior to the development of the modern tank designs should consider removing these tanks inasmuch as leak detection systems are far from foolproof.

Monitoring wells offer a secondary method of groundwater protection. Groundwater monitoring wells can be installed down to the groundwater level, or, in some cases, a device called a lysometer can be used to detect contamination at a given elevation above the groundwater. Any monitoring well test program should be designed thoughtfully to monitor for the correct contaminants and avoid excessive analytical costs. The frequency and duration of testing are considerations in designing groundwater monitoring plans. Environmental agencies typically ask facilities to prepare monitoring plans that include a great number of samples taken over a long period of time. A common compromise with the agencies is to reduce the frequency of testing once an acceptable level of groundwater quality has been achieved.

16.5 TESTING REQUIREMENTS

What are you required to test in groundwater? The answer depends on the purpose of the testing. Most states have established numerical groundwater quality standards that vary by contaminant and by use. These standards can be obtained from the health department and environmental agency in your area. Groundwater quality standards are normally related to providing water suitable for drinking and include parameters for inorganics, organics, volatile organics, pH, radionuclides, color, and/or odor. Standard Environmental Protection Agency (EPA) test methods are available for instructing a test lab on what contaminants and levels of detection are to be examined.

In the case of site remediation, some states have created a separate set of water quality requirements. These standards typically recognize the realities of groundwater remediation and are designed to allow for possible future use of the site. Even these standards are subject to financial analysis and risk evaluation prior to beginning remediation, with the highest beneficial use of the property and the intended use (if any) of the groundwater subject to careful consideration. An example of state-approved groundwater cleanup

Table 16.1 Washington State—Method A cleanup levels (groundwater)

Hazardous Substance	CAS Number[a]	Cleanup Level[b]
Arsenic	7440-38-2	5.0 μg/liter
Benzene	71-43-2	5.0 μg/liter
Cadmium	7440-43-9	5.0 μg/liter
Chromium (total)	7440-47-3	50.0 μg/liter
DDT	50-29-3	0.1 μg/liter
1,2 Dichloroethane	107-06-2	5.0 μg/liter
Ethylbenzene	100-41-4	30.0 μg/liter
Ethylene dibromide	106-93-4	0.01 μg/liter
Gross alpha particle activity		15.0 pCi/liter
Gross beta particle activity		4.0 mrem/year
Lead	7439-92-1	5.0 μg/liter
Lindane	58-89-9	0.2 μg/liter
Methylene chloride	75-09-2	5.0 μg/liter
Mercury	7439-97-6	2.0 μg/liter
PAH (polyaromatic hydrocarbons)		0.1 μg/liter
PCB mixtures		0.1 μg/liter
Radium 226 and 228		5.0 pCi/liter
Radium 226		3.0 pCi/liter
Tetrachloroethylene	127-18-4	5.0 μg/liter
Toluene	108-88-3	40.0 μg/liter
Total petroleum hydrocarbons		1000.0 μg/liter
1,1,1 Trichloroethane	71-55-6	200.0 μg/liter
Trichloroethylene	79-01-5	5.0 μg/liter
Vinyl chloride	75-01-4	0.2 μg/liter
Xylenes	1330-20-7	10.0 μg/liter

[a] Chemical Abstract Service Number
[b] See regulation for qualifiers.

levels derived from Washington State's Model Toxics Control Act is shown in Table 16.1.

16.6 REMEDIATION

There are several ways in which an environmental manager can become involved in a groundwater remediation project. For example, a spill at your site could impact groundwater. A spill or leak at a property in the vicinity of your site may impact the groundwater beneath your facility's property, or disposal of hazardous waste from your facility in a site owned and operated by a third party may result in a groundwater remediation project involving your facility as a potentially responsible party. Whatever the reason, the remediation of groundwater can be technically challenging and costly. Major groundwater remediation projects usually involve lengthy negotiations with the environmental agencies to establish acceptable cleanup levels and remediation techniques. Although there are many technology options for groundwater remediation, the technologies generally fall into the following three categories:

- Pumping the groundwater and treating it in above-ground facilities.
- Treating the groundwater in place biologically with air, oxygen, or precipitants.
- Preventing or impeding the spread of the groundwater contamination (allows for natural attenuation).

Whichever option is selected, a health risk assessment will likely be required by the environmental agency to determine the risks to human health and the environment associated with the existing and proposed levels of contamination. The subject of risk assessment is discussed in some detail in Chapter 24 of this book.

If the selected technology is to pump and treat the contaminated water, there must be a new discharge of the treated groundwater. If the treated water is meeting groundwater standards, the treated water may be reinjected back into the groundwater. However, if this approach is not possible, the design of a pumping and piping system to allow discharge through a municipal or industrial treatment system may be required. In that case, it will also be necessary to obtain a permit to discharge this effluent through either a direct discharge or to a publicly owned treatment works (POTW). Because pump and treat systems often require many years of operation to reach acceptable levels of groundwater contamination, environmental managers are often asked to assume responsibility for the ongoing operation and maintenance of the system.

The second technical approach to remediating groundwater is to treat the contamination in place. This approach has gained popularity in recent years as it usually offers cost-effective treatment and eliminates the need to manage the discharge of contaminants associated with pumping and treating. In-place treatment is typically accomplished by injecting an agent into the contaminated groundwater to separate the contaminants from the groundwater and surrounding aquifer, change the contaminants into a less toxic or less mobile form, or degrade them to eliminate their toxicity. Of course, in-place treatment strategies must be evaluated on a site-specific basis and approved by the environmental agency overseeing the project.

Another approach is to leave the contaminated groundwater in place if scientific evidence is adequate to show that the plume is confined, spreading extremely slowly, or not likely to cause significant additional harm to adjacent property or uses. The decision to leave contaminated groundwater in place typically must be supported by a detailed risk assessment demonstrating that there is no significant risk to human health or the environment. Groundwater contamination that is left in place is usually subject to a variety of ''institutional'' controls. For example, the local government may prohibit use of groundwater from the contaminated area as drinking water or rezone the property to preclude residential use. Although this approach may save remediation costs, it will result in the loss of property value and may subject the facility to damage claims from adjacent landowners. Certainly, all of these factors must be considered in selecting the most appropriate groundwater remediation strategy for a site.

17

WATER RIGHTS

Because water rights are often administered by state environmental agencies, environmental managers are typically involved in the acquisition and permitting of water rights. The subject of water law is essentially a study of property ownership concepts. As population and associated development have increased and groundwater aquifers have been contaminated by environmental pollution, water is becoming a more precious commodity. Proper management and acquisition of the legal rights to water are absolutely essential to most industrial facilities. Volumes of information are available regarding the creation and administration of water rights. Our purpose in this chapter is to acquaint the environmental manager with the basic principles of water law and explain how water rights relate to managing environmental compliance at an industrial facility.

17.1 WATER RIGHTS SYSTEMS

There are three different systems for administering rights in the United States: riparian, prior appropriation, and hybrid systems. Only two states have developed systems outside these approaches. The important aspects of each system are summarized in the following paragraphs.

17.1.1 Riparian Rights

Riparian rights are water rights that belong to landowners by virtue of their owning property adjacent to a waterway, such as a river. The riparian doctrine applies in one form or another in 29 states, which are mostly east of the Mississippi River. The riparian rule allows property owners with riparian access to use water in a way that is ''reasonable'' relative to the demands of all other users. If there is insufficient water to satisfy the reasonable needs of all riparian users, they all must reduce their usage of water in proportion to their rights. Riparian rules have been altered by statute in most states to require riparian users to obtain permits from the authorized state agency before water can be used.

17.1.2 Prior Appropriation

The prior appropriation doctrine is the basis for water rights in nine states of the arid West. The prior appropriation water right is based on the principle of ''first in time, first in right'' and provides that anyone with a prior recognized water right (a ''senior'' water right) has superiority over someone who later began using the water (a ''junior'' right). Once a person puts a designated amount of water to a beneficial use and complies with specified administrative and permit requirements, the water right is deemed to be ''perfected,'' or legally valid, and remains valid as long as it continues to be used and water is available. In drought years it is common for junior rights to be curtailed or eliminated for periods of time in deference to a senior water right.

17.1.3 Hybrid Systems

Ten states have hybrid water rights systems, including California and Texas. These states originally followed the riparian doctrine but later converted to a prior appropriation system that recognized existing riparian rights.

17.1.4 Other Systems

Hawaii and Louisiana have water rights programs that do not fit neatly within any of the three categories. When working in these states, it is important to consult local experts before deciding how to proceed on any water rights issue.

17.2 PERMIT PROGRAMS

Most states now have some sort of permit system for surface water rights, which is usually administered by a state environmental agency. Several states also require permits for the use of groundwater. These permit programs require anyone wanting to divert or impound water to obtain a permit from the authorized agency. This permitting effort typically involves an established review process. Applicants are generally required to submit a formal written application, which is subject to public notice and the possibility of a public hearing. The permitting authority is then required to evaluate certain criteria to determine whether the issuance of the water right permit is in the public interest. Permits are usually specific as to the location, volume, and rate of diversion and the location and nature of the permitted use. Permittees are generally required to monitor and report their water use and pay a modest administrative fee.

About half of the permit states grant a perpetual permit. In the others, the permit is issued for a fixed term ranging from 10 to 50 years. Permits are generally forfeited if their use does not begin soon after they are granted or if use is suspended for longer than a fixed period of time.

17.3 WATER RIGHTS MANAGEMENT

17.3.1 Establishment of Water Rights

When process water is available from a municipal source, acquiring a water right may not be required. However, for large-quantity water users, the cost of buying municipal water can negatively impact the financial success of the facility, thus creating a financial incentive to obtain an independent water right.

The first step in managing a facility's water rights is to verify that legitimate water rights exist and that the rights are certified and properly documented. The water rights must also be adequate to meet both the instantaneous and average current use of water by the facility. An example of a water rights certificate is shown in Figure 17.1.

If a new facility or facility modification requiring additional water use is proposed, it is necessary to acquire properly documented water rights before committing to the new project. The process of obtaining a certified water right is time-consuming, and, occasionally, the necessary water rights are not available. Under the provisions of contemporary water law, facilities are no longer allowed to drill groundwater production wells or pump from a nearby surface water supply without first obtaining the proper permits. Failure to follow these procedures can subject a facility to both administrative penalties and lawsuits by neighboring water users.

17.3.2 Documenting Use

Documentation of water use from groundwater wells and surface waters is essential in maintaining water rights. Water rights are often defined by two

STATE OF OREGON

COUNTY OF MARION

CERTIFICATE OF WATER RIGHT

This is to Certify, That JOHN HANCOCK *of* Box 100, Salem, *State of* Oregon, 97309, *has made proof to the satisfaction of the STATE ENGINEER of Oregon, of a right to the use of the waters of* XY River, *a tributary of* XYZ River, *for the purpose of* irrigation *under Permit No.* 007 *of the State Engineer, and that said right to the use of said waters has been perfected in accordance with the laws of Oregon; that the priority of the right hereby confirmed dates from* September 10, 1953, *that the amount of water to which such right is entitled and hereby confirmed, for the purposes aforesaid, is limited to an amount actually beneficially used for said purposes, and shall not exceed* 1.50 cubic feet per second, *or its equivalent in case of rotation, measured at the point of diversion from the stream. The point of diversion is located in the* SW ¼ SE ¼, as projected within Smith DLC #60, Section 8, Township 7 South, Range 2 East, W.M.

The amount of water used for irrigation, together with the amount secured under any other right existing for the same lands, shall be limited to one-fiftieth *of one cubic foot per second per acre,* or its equivalent for each acre irrigated and shall be further limited to a diversion of not to exceed 5 acre feet per acre for each acre irrigated during the irrigation season of each year, *and shall conform to such reasonable rotation system as may be ordered by the proper state officer.*

A description of the place of use under the right hereby confirmed, and to which such right is appurtenant, is as follows:

2.8 acres in the NE ¼ SE ¼
3.46 acres in the NW ¼ SE ¼
10.6 acres in the SW ¼ SE ¼
12.0 acres in the SE ¼ SE1/4
All as projected within Smith DLC #60
Section 8
Township 7 South, Range 2 East, W.M.

The right to the use of the water for the purposes aforesaid is restricted to the lands or place of use herein described.

WITNESS the signature of the State Engineer, affixed

this 12th day of September, 1955.

ANN B. LEWIS
State Engineer

Recorded in State Record of Water Right Certificates, Volume 17, *page* 23000.

Figure 17.1 Certificate of Water Right

parameters established in the water permit. First, an extraction rate such as 1,000 gallons per minute and, second, an annual total use amount, which is usually described as a certain number of acre-feet, A facility can lose its water right or part of the right if the use is terminated or if the full right is not exercised. Therefore, the facility should maintain records for water use that include flow rates for both annual and instantaneous peak uses.

The best method for documenting water use is direct measurement of the source of the water supply. In some facilities this may mean monitoring several stream flows. Flow should be measured in such a way that it is easy to calculate both the peak and annual water use of the facility. If direct flow measures are not available, surrogates such as well pump motor amperage and pump discharge pressure recordings, used in conjunction with pump design flow curves, can be used to provide a fairly accurate estimate of water use.

17.3.3 Excess Water Rights

Some facilities may conclude that their current water rights exceed their normal or anticipated use or that the water rights are no longer needed. If a water right has not been relinquished under state or local law, the facility owner may be able to sell or transfer the water right to a third party. In general, a water right that is transferred cannot be used for a purpose other than the type of use specified in that water right. Moreover, the proposed transfer must not injure other water uses. The transfer and sale of water rights is a complicated area of water law, and facilities contemplating the sale of a water right should consult a qualified water rights expert familiar with state and local requirements.

18

SOLID AND HAZARDOUS WASTE MANAGEMENT

Statute:	Resource Conservation and Recovery Act	42 USC § 6901 et seq.
Regulations:	Solid Waste	40 CFR Parts 240 to 258
	Hazardous Waste	40 CFR Parts 260 to 282

Solid and hazardous waste management is an aspect of environmental protection that has received relatively little federal attention until recently. Comprehensive federal regulation of solid waste management began with the enactment of the Resource Conservation and Recovery Act of 1976 (RCRA), which amended the Solid Waste Disposal Act. In general terms, RCRA applies to the universe of materials defined in the Act as "solid waste." RCRA establishes procedures for the management, reuse, or recovery of solid waste and imposes detailed requirements governing the highly regulated subcategory of solid waste determined to be "hazardous."

18.1 FEDERAL REGULATION OF SOLID WASTE

Subtitle D of RCRA contains general authority for the Environmental Protection Agency (EPA) to enact a solid waste regulatory program. RCRA's objective for solid waste management is outlined as follows:

> To assist in developing and encouraging methods for the disposal of solid waste which are environmentally sound and which maximize the utilization of

valuable resources including energy and materials which are recoverable from solid waste and to encourage resource conservation (42 U.S.C. § 6941).

EPA has promulgated regulations that establish guidelines for managing various types of solid wastes and developing and implementing state solid waste management plans. EPA has also developed criteria governing municipal solid waste landfills. These rules contain provisions regarding landfill location, operating criteria, design criteria, groundwater monitoring and corrective action, closure and postclosure care, and financial assurance criteria.

Most states are experiencing a growing shortage of environmentally acceptable solid waste disposal sites, a rising standard of what is environmentally acceptable, and increased difficulty in siting disposal facilities. There also are a variety of public concerns relating to groundwater contamination and other environmental problems associated with past solid waste disposal practices, which have resulted in the closure of many older municipal landfills. Given the urgency of these problems and the absence of clear federal leadership, the regulatory requirements governing the management of solid waste are, for the most part, being developed and implemented by the states. States, in turn, delegate much of the solid waste management authority to municipal or county agencies.

Because states are summarily responsible for administering solid waste programs, this discussion will focus primarily on some of the key concepts of solid waste management and summarize a sampling of provisions typically found in state solid waste management programs.

18.2 STATE SOLID WASTE MANAGEMENT

In most states, the state environmental regulatory agency regulates the management of solid waste. Local governments, through local codes, land use regulations, and comprehensive plans, also play a key role in solid waste management and the siting of solid waste disposal facilities. The state usually retains final authority and may override local decisions in extraordinary circumstances. State solid waste policies strongly encourage waste reduction, reuse, recycling, and energy recovery. Only waste that cannot be reduced, reused, or recovered should be disposed of in landfills or other facilities.

State solid waste management statutes and regulations typically establish requirements for permitting solid waste disposal sites. Siting a new landfill involves analyzing the scientific, logistical, and societal factors associated with each alternative location and can be a costly and time-consuming process. Once a location is selected and the landfill is constructed in accordance with the specified design factors, a landfill operating permit is issued. Requirements imposed by solid waste disposal site operating permits include the following:

- Use of ''best management practices'' to prevent contamination of the surrounding environment,
- Groundwater monitoring and corrective action if necessary,
- Recycling procedures,
- Vector and bird control,
- Quarterly reporting,
- Gas emissions monitoring and control, and
- Closure and postclosure plans and financial assurance.

Most states also establish some form of streamlined permit program for facilities that have a short-term waste storage need but do not plan to become a permanent storage or disposal site. Permits of this nature are valid for a relatively short period of time (e.g., six months) and contain abbreviated monitoring and reporting requirements.

State solid waste legislation also regulates the handling and disposal of infectious wastes, household wastes, plastic bags, used batteries, waste tires, and other specific solid waste streams.

18.3 FEDERAL REGULATION OF HAZARDOUS WASTES

Managing hazardous waste is one of the most challenging aspects of environmental management. RCRA, Subtitle C, is the major regulatory program governing the management of solid wastes that are defined as hazardous. As required by RCRA, EPA has developed a ''cradle-to-grave'' regulatory scheme for managing hazardous wastes and controlling their ultimate disposal. EPA's RCRA program establishes standards for hazardous waste management applicable to hazardous waste generators, transporters, and treatment, storage, and disposal (TSD) facilities. The hazardous waste regulations are considered by most experts to be the most complicated federal environmental regulatory program. Several states have enacted hazardous waste legislation even more stringent than the federal law.

One of the most important aspects of managing solid and hazardous wastes is determining what wastes must be classified as hazardous. Under the hazardous waste regulations, a material must first fit the definition of a ''solid waste'' before it can be considered a hazardous waste. Solid waste is generally defined as a solid, liquid, or gas that is a discarded material and is not expressly excluded from the definition of solid waste or granted a variance from its classification as a solid waste. The regulations categorically exclude certain materials from the definition of solid waste, including domestic sewage, point source discharges, and irrigation return flows. Based on this regulatory definition, the key to determining whether a material is a solid waste is to establish whether the material is discarded. EPA's definition of ''discarded material'' includes any material that is abandoned, inherently waste-like, or recycled in certain ways. In regard to recycled materials, the regulations state that material is recycled if it is used, reused, or reclaimed. EPA has established a detailed matrix in the hazardous waste regulations that can be used to determine whether recycled materials are solid wastes. Once it is established that a material is a ''solid waste,'' there are two ways in which it is designated a hazardous waste: it is specifically listed or it exhibits one of four hazardous characteristics.

18.3.1 Listed Hazardous Wastes

There are three categories of listed hazardous waste:

1. *Hazardous wastes from nonspecific sources.* Hazardous wastes from nonspecific sources are generic wastes that may be generated by any number of industrial processes. These wastes are listed at 40 CFR § 261.31. An example of hazardous wastes from a nonspecific source are spent halogenated solvents used in degreasing, such as tetrachlorethylene, trichloroethylene, methylene chloride 1.1.1.-trichloroethane, carbon tetrachloride, and chlorinated fluorocarbons.
2. *Hazardous wastes from specific sources.* Hazardous wastes from specific sources are hazardous wastes generated by processes in a particular industry. An example of hazardous wastes from a specific source are tank bottoms (leaded) in the petroleum refining industry. Hazardous wastes from specific sources are listed at 40 CFR § 261.32.
3. *Specified commercial chemical product wastes.* EPA has also listed at 40 CFR § 261.33 commercial chemical products that are wastes if and when they are discarded or intended to be discarded. This list contains hundreds of chemical products identified either by trade or by chemical name. An example of a chemical product in this list is pentachlorophenol.

EPA's lists of hazardous wastes are subject to change, and waste generators are responsible for regularly reviewing the lists to determine whether their wastes have been listed.

Hazardous waste–generating facilities should also ensure that listed hazardous wastes are not mixed with nonhazardous solid wastes inasmuch as, subject to certain exceptions, mixtures of those wastes are subject to hazardous waste regulation under the rationale of EPA's ''mixture rule.'' EPA has also developed another rule relating to listed hazardous wastes—the ''derived from'' rule—which provides that any material derived from the treat-

Material:
Origin:
Destination:
Description of Storage and Handling System:
Listing:
Hazardous Wastes from Nonspecific Sources:
Hazardous Wastes from Specific Sources:
Specified Commercial Chemical Product Wastes:
Acutely Hazardous Wastes:
Tests:
Toxicity Characteristic Leaching Procedure (TCLP):
Corrosivity:
Ignitability:
Reactivity:
Other Comments:

Figure 18.1 Waste stream determination work sheet

ment, storage, or disposal of a listed hazardous waste remains a hazardous waste until that material is delisted.

18.3.2 Hazardous Characteristics

In addition to the listed hazardous wastes, there are other solid wastes that can be considered hazardous if a ''representative sample of the waste'' exhibits any of the following four characteristics: ignitability, corrosivity, reactivity, or toxicity.

The hazardous waste regulations put the burden on the waste generator to determine whether its waste exhibits one or more of the hazardous characteristics. The regulations establish detailed criteria for each characteristic and, in many cases, specify the test procedures to be used.

In order to ensure compliance with the hazardous waste regulations, generators should evaluate each waste stream at a facility and determine, either through testing or general knowledge, whether the waste is hazardous. Additional evaluation is required if changes in a waste stream occur or new chemicals are introduced to the process. A sample waste stream determination work sheet is provided in Figure 18.1.

18.4 GENERAL STANDARDS APPLICABLE TO HAZARDOUS WASTE GENERATORS

Environmental managers of facilities that generate hazardous wastes must ensure that these wastes are properly handled and disposed of. The hazardous waste regulations define "generator" as any person, by site, whose act or process produces hazardous waste. Under the regulations, generators of hazardous wastes must obtain an EPA identification number and prepare a manifest for each shipment of hazardous wastes sent off-site for treatment, storage, or disposal at an EPA-licensed facility. The manifest system is designed to track hazardous waste from the generator through the transporter to the disposal facility. The disposal facility is required to return a signed copy of the manifest to the generator. Generators must also package and containerize the hazardous waste and ensure that the waste is transported in accordance with Department of Transportation regulations for hazardous materials.

Generators of hazardous waste are required to familiarize all employees with proper waste-handling and emergency procedures relevant to their responsibilities within the facility. Generators must also designate an employee as the emergency coordinator who will be responsible for coordinating all emergency response procedures.

All hazardous wastes, whether treated and disposed of on-site or sent off-site to an RCRA treatment, storage, or disposal facility, are subject to testing and record-keeping requirements. Generator responsibilities include determining whether the waste is subject to the Land Disposal Restriction (LDR) rules, which prohibit the land disposal of many hazardous wastes, determining what constituent levels are in the waste, determining the applicable treatment standards, and certifying that waste sent to a disposal facility meets the selected treatment standard.

EPA has also promulgated treatment standards for hazardous debris. Hazardous debris is not only that generated from hazardous substance remediation at a facility but can also be that associated with demolition or decommissioning activities. In general, debris that contains listed hazardous waste or exhibits a hazardous characteristic must be managed as a hazardous waste. EPA's hazardous debris rule establishes 17 different treatment technologies for contaminated debris. Environmental managers should become familiar with this rule and consider its application whenever a facility is engaged in any demolition or decommissioning activities.

18.5 HAZARDOUS WASTE GENERATOR CATEGORIES

There are three categories of hazardous waste generators defined by EPA. These categories and the applicable requirements are summarized in the following paragraphs.

1. *Conditionally Exempt Small-Quantity Generator*. If a facility does not generate more than 100 kilograms (220 pounds or approximately 25

gallons) of hazardous waste in a month, the facility is classified as a "conditionally exempt" small-quantity generator and is not subject to many of the hazardous waste management and permitting requirements. In general, conditionally exempt small-quantity generators are required to identify all hazardous waste generated and send the waste to an EPA-permitted hazardous waste facility, using a uniform hazardous waste manifest.

2. *100–1,000 Kilograms/Month Small-Quantity Generator.* A facility that generates more than 100 but less than 1,000 kilograms of hazardous waste in a month is required to comply with most of the hazardous waste management regulations applicable to generators. In this generator category, hazardous waste may be accumulated on-site for up to 180 days (or up to 270 days if the waste has to be transported over a distance of 200 miles or more) without a permit, provided the waste accumulated on-site never exceeds 6,000 kilograms. This category of generator is also subject to reduced record-keeping and reporting requirements.
3. 1,000 or More Kilograms/Month Generator. Facilities that generate 1,000 kilograms or more of hazardous waste in one month must comply with all of the hazardous waste management requirements applicable to generators. Generators of more than 1 kilogram per calendar month of wastes that have been designated as "acutely hazardous" also are subject to all of the regulations that apply to this category of generator. Generators of hazardous waste in this category are required to obtain a hazardous waste storage permit if they store their waste for longer than 90 days. Because the permitting requirements for hazardous waste TSD facilities are extremely onerous, environmental managers should develop a foolproof system to ensure that hazardous wastes are never stored longer than 90 days.

18.6 HAZARDOUS WASTE MANAGEMENT PRACTICES

Facilities that generate large quantities of hazardous waste are subject to a broad array of regulatory requirements. Some of the common errors found at facilities generating and handling large quantities of hazardous wastes are summarized in Table 18.1.

Of course, hazardous waste generators should always focus on waste minimization techniques and the use of raw materials in the process that will produce a nonhazardous waste stream. Additional thoughts regarding waste minimization and pollution prevention are found later in this chapter and in Chapter 30.

Table 18.1 Common hazardous waste management violations

WASTE ANALYSIS
Failure to conduct a waste stream determination
Failure to use proper sampling or analytical methods
Failure to use proper units of measure
DRUM MANAGEMENT
Improper labeling
No accumulation start date marked on the drum
Insufficient aisle space between drums
Failure to keep containers closed
RECORD KEEPING AND REPORTING
Failure to submit annual or biennial report
Failure to use proper agency forms
Failure to update contingency plan
Failure to submit exception report for missing manifests
Failure to keep waste manifests for three years
INSPECTIONS
Failure to inspect emergency response equipment on a regular basis
Failure to conduct equipment inspections
Failure to document equipment inspections
Failure to document corrections made to equipment

18.7 STANDARDS FOR HAZARDOUS WASTE TREATMENT, STORAGE, AND DISPOSAL FACILITIES

Hazardous waste treatment, storage, and disposal facilities are subject to the most stringent set of requirements established by the hazardous waste regulations. In order to operate a facility that treats, stores, or disposes of hazardous wastes, the owner of a facility must obtain an RCRA operating permit and comply with the detailed design and operating standards established by Part 264 of the hazardous waste regulations.

Facilities that treat, store, or dispose of hazardous wastes must comply with a comprehensive set of regulations requiring personnel training, emergency response programs, inspection and reporting responsibilities, and other standards designed to protect public health and the environment. The regulations also require the owner of a TSD facility to provide appropriate financial assurance for closure and postclosure costs of the facility.

A TSD facility is also subject to the Land Disposal Restrictions and must comply with established record-keeping requirements and ensure that all applicable restricted wastes are properly treated and disposed of.

EPA has authority under RCRA to require owners of hazardous waste treatment, storage, and disposal facilities to take "corrective action" to address releases of contaminants from their facilities. In appropriate cases, EPA often uses RCRA corrective action rather than the authority of the Superfund to require facility owners to remediate releases of hazardous substances, inasmuch as the RCRA permitting process allows EPA to require corrective action as a condition of obtaining or continuing an operating permit for the TSD facility.

18.8 USED OIL

After years of deliberation, EPA decided in 1992 not to list used oil as a hazardous waste, regardless of whether the used oil is disposed of or recycled. However, EPA did establish standards for used oil management. The regulations establish standards for the following categories of used oil handlers:

- Used Oil Generators
- Used Oil Collection Centers and Aggregation Points
- Used Oil Transporter and Transfer Facilities
- Used Oil Processors and Re-Refiners
- Used Oil Burners Who Burn off Specification Used Oil for Energy Recovery
- Used Oil Fuel Marketers

The rules also govern the disposal of used oil and explain the very limited circumstances in which used oil can be used as a dust suppressant. Most industrial facilities are subject only to the regulations governing used oil generators. These rules establish standards for used oil storage, labeling, and spill response. Used oil that exhibits any of the characteristics of hazardous waste must be managed as a hazardous waste under RCRA. Nonhazardous used oil can be managed in a variety of ways, although most used oil is either recycled or burned for energy recovery.

18.9 STATE HAZARDOUS WASTE PROGRAMS

The federal hazardous waste law applies in all states, but each state may adopt its own hazardous waste program and that program may be more stringent than the federal rules. Although most states incorporate the federal hazardous waste rules by reference with little or no modification, several states have enacted hazardous waste management rules that are quite different from the federal law. Upon approval of a state program by EPA, the state is authorized to regulate hazardous wastes in lieu of EPA.

Environmental managers should always know whether the state has been delegated authority to administer the hazardous waste program and whether the state program is different from the federal rules. A manager should never assume that compliance with the federal rules ensures compliance with state laws, because, as mentioned earlier, states frequently develop their own programs.

18.10 SOLID WASTE MINIMIZATION AND RECYCLING

Facility environmental managers should be the champions of solid and hazardous waste management and continuously focus on opportunities for waste minimization and recycling. Among the key elements of a solid waste minimization program are the following:

- Reduce the amount of raw material required to manufacture the product.
- Purchase products that contain less packaging or come in refillable containers.
- Use products that are more durable, repairable, or reusable.
- Reuse the material/product a number of times.
- Consider modifying the process and equipment design to reduce the amount of waste generated.
- Reuse materials directly in the manufacturing process.
- Institute new procedures, such as preventive maintenance, to reduce waste generation.
- Minimize the quantities of raw material on-site to eliminate surplus that could become waste if the product is changed or discontinued.
- Closely police the introduction of any new products into the facility.
- Discourage vendors from leaving large quantities of "sample" products for trial use, or establish a policy that unused samples should be returned to the vendor.

In addition to these waste minimization procedures, facilities should also develop comprehensive programs to segregate waste streams, reduce the volume of waste handled, and facilitate reuse and recycling. Employee education is the key to a successful waste segregation program, and environmental managers should devise training programs and systems that encourage and reinforce the importance of waste segregation and recycling. Once a company has a waste segregation program in place, the environmental manager should explore various ways of recovering, reusing, or recycling the material either within the company or through outside vendors.

Given the high cost associated with solid waste handling and disposal, a strong waste minimization and recycling program can benefit both the environment and the economic performance of a facility. A more detailed discussion of waste minimization programs can be found in Chapter 30.

18.11 SELECTING A HAZARDOUS WASTE TREATMENT, STORAGE, AND DISPOSAL FACILITY

The environmental manager's job does not end once shipments of hazardous waste have left the site. Under the provisions of the Superfund Law, generators of hazardous waste can be held liable if the hazardous waste TSD facility mismanages the waste in a manner that results in the release of hazardous substances to the environment. In light of this liability, environmental managers should consider the following factors when selecting a TSD facility.

18.11.1 Compliance History

Before selecting a hazardous waste management facility, the environmental manager should consult with applicable federal, state, or local environmental authorities to learn the regulatory history of the management facility.

In addition to verifying the permitted status of the facility, the environmental manager should also determine whether enforcement proceedings have ever been initiated against the facility and review reports of any hazardous waste inspections conducted at the site. Identification of past compliance problems or pending agency action should alert you to possible problems associated with the use of a management facility. It is unlikely that any disposal facility will possess an unblemished regulatory record. Therefore, it is essential to exercise good judgment regarding the significance of a site's regulatory history.

18.11.2 On-Site Inspection

An on-site inspection is an important prerequisite to selecting a hazardous waste management facility. Before a waste generator selects a particular TSD facility, the environmental manager should meet with the operators of the facility to determine its compliance with various regulatory requirements.

This determination should include an examination of the facility's operating permits and a review of the facility's liability insurance coverage for claims arising from sudden and accidental occurrences related to the handling of hazardous waste. It is also important to determine whether the facility has the financial assurances necessary to ensure that funds will be available for properly closing the facility and for maintaining and monitoring it after closure.

While at the site, the waste generator should observe the facility's physical plant and obtain a complete description of the facility's waste handling procedures. The facility's record keeping and manifest procedures should also be reviewed to determine whether the facility clearly understands the applicable regulatory requirements. Finally, an on-site inspection of the facility affords the environmental manager a good opportunity to discuss with facility operators the history and background of the operation and plans for future waste handling activities. If, after this on site inspection, you conclude that a facility is deficient in its record keeping and operating procedures, serious consideration should be given to selecting another facility.

18.11.3 Financial Analysis

Because the possibility of a waste generator's liability for improper hazardous waste disposal is closely connected to the financial accountability of the TSD facility, the company should carefully examine the management facility's financial statement.

Generally speaking, waste generators should choose to have their hazardous waste handled by the disposal facility with the strongest financial picture. Along these same lines, national waste management companies usually offer more protection to waste generators than small, regional businesses.

18.11.4 Review of Customer List

The environmental manager's examination of a hazardous waste management facility should include a review of the facility's list of customers and the nature and quantity of wastes contributed to the site by each of the customers. Because liability for improper disposal of hazardous waste is usually shared among parties who contribute waste to the disposal site, waste generators are well-advised to dispose of their hazardous waste at facilities used by large, financially responsible companies. A waste generator's potential liability for improper hazardous waste disposal increases dramatically when that waste generator is the largest company contributing waste to a disposal site or when it contributes a relatively large amount of waste to the site.

18.11.5 Proper Management Methods

Much of today's hazardous waste contamination was caused by improper land disposal of hazardous waste. To the extent that a waste generator can dispose of its waste through incineration, the risk of future liability is significantly reduced. However, selection of incineration as the preferred disposal method is recommended only when the incineration facility is permitted and efficiently operated.

18.11.6 Waste Management Contracts

Hazardous waste should never be delivered to a hazardous waste management facility without a written contract with the facility. Proposed contracts with hazardous waste management facilities should be carefully reviewed by legal counsel.

18.11.7 Periodic Review of the Management Facility

Once a hazardous waste management facility has been selected, the environmental manager should periodically review the compliance status of the facility. This periodic review should include consultation with regulatory officials and on-site inspections to ensure that the facility is maintaining its operating procedures and compliance standards.

If a detailed review is not possible, the waste generator should at least visit the facility to observe its operating procedures, consult with regulatory authorities to review the facility's compliance history, and review any hazardous waste management contract with legal counsel.

Although adherence to these recommendations does not totally protect a waste generator from liability for a management facility's improper disposal of hazardous waste, it does provide a practical method of selecting the best available hazardous waste management facility. Given the current status of the Superfund and the hazardous waste regulatory program, there is no such thing as "risk-free" hazardous waste disposal.

18.12 UNDERGROUND STORAGE TANKS

Under Subtitle I of RCRA, Congress directed EPA to establish a regulatory program that would prevent, detect, and clean up releases from underground storage tanks (USTs) containing petroleum or hazardous substances. By December 22, 1998, all tanks were required to meet stringent federal and state standards for underground storage tanks. Tanks that were installed before December 1988 were required to be replaced or retrofitted by December 22,

1998, to meet the new standards. New tanks must meet the new standards at the time of installation. The criteria for new installations of underground storage tanks are complex and govern the following five subject areas:

- Design and installation
- Operating requirements
- Corrosion protection
- Leak detection
- Corrective measures

The provisions of 40 CFR Part 280 and applicable state regulations require owners or operators of regulated underground storage tanks to obtain a UST permit. In general, a UST permit is required for an existing or new tank that:

- Stores petroleum products, used motor oil, or hazardous substances such as industrial solvents, pesticides, and herbicides,
- Is larger than 110 gallons, and
- Has 10% or more of the total volume, including piping, beneath the ground.

The regulations identify several categories of USTs, such as heating oil tanks, and septic tanks, that are exempt from the permit requirement.

UST permits typically contain requirements that direct UST owners to use only licensed UST service providers for installation and maintenance of the tank, monitor the tank for leaks and releases, and report all suspected or confirmed releases to the appropriate regulatory agency. State law usually requires contractors who install, retrofit, decommission, or test USTs to obtain a license to work on underground storage tanks. State law also establishes standards for treating petroleum-contaminated soil from a UST release.

Separate regulations were issued by EPA for financial responsibility requirements for tank owners and operators. Financial assurance, available through various mechanisms such as liability insurance, self-insurance tests, guarantees, trust funds, and letters of credit, must be established for a per-occurrence liability coverage of $500,000 or $1 million, depending on the amount of petroleum handled at a facility per month. An annual aggregate coverage of $1 or $2 million must also be provided, depending on the total number of tanks that are subject to the regulations.

With the implementation of the UST program, USTs are increasingly being closed and replaced with aboveground tanks. However, because USTs still exist at some facilities and because aboveground tanks are not practical in some situations, the UST regulatory program remains an important environmental management issue for many facilities.

19

SUPERFUND LIABILITY

Statute:	42 USC §§ 9601 to 9675
Regulations:	40 CFR Parts 300 to 311

In December 1980, Congress passed the Comprehensive Environmental Response, Compensation, and Liability Act of 1980. Commonly known as CERCLA or "Superfund," the Act provides authority and funding for cleaning up inactive hazardous waste disposal sites, spills, and other discharges of hazardous substances into the environment.

The first several years of Superfund were marked by a wave of litigation regarding the implementation and constitutionality of the statute. Consequently, relatively little cleanup work was actually accomplished. As a result, Congress enacted the Superfund Amendments and Reauthorization Act (SARA) in 1986. In addition to correcting some of the practical flaws in the 1980 Superfund Law, SARA established new and independent regulatory programs such as the Emergency Planning and Community Right-to-Know Act (SARA Title III). Much of the law governing the administration of Superfund is found in court decisions and a variety of published and unpublished Environmental Protection Agency (EPA) policy statements. The purpose of this chapter is not to provide a detailed discussion of Superfund and how its provisions have been interpreted by the courts, but rather to introduce environmental managers to some of the key concepts of the Superfund program.

19.1 SPILLS AND DISCHARGES OF HAZARDOUS SUBSTANCES

Superfund requires that any person in charge of a site where hazardous materials are located must notify EPA as soon as he or she is aware of an unpermitted release, discharge, or spill of a ''hazardous substance'' that is greater than or equal to the reportable quantity for that substance. Notifications to EPA under this provision of the law are made to the National Response Center at U.S. Coast Guard headquarters in Washington, D.C. The purpose of the notification requirement is to ensure that the appropriate federal government officials are informed of the release in a timely manner, to enable them to determine whether federal response action is required. The list of the materials identified by EPA as hazardous substances and the reportable quantity for each substance are published at 40 CFR Part 302. Armed with this information, the environmental manager should prepare a list of the hazardous substances used at the facility and the applicable reportable quantities. This information will allow the manager to make a prompt decision as to whether to notify EPA of a release. There are several important exclusions from this release notification requirement. For example, petroleum is excluded from the Superfund definition of hazardous substance, although petroleum discharges to ''navigable waters'' must be reported to the National Response Center under the authority of the Clean Water Act. There also are exclusions from Superfund spill reporting requirements for engine exhaust emissions, releases from consumer products in consumer use, and certain ''federally permitted releases.'' Chapter 25 of this book discusses the reporting requirements of CERCLA and other environmental statutes in further detail.

It is important to note that release notifications to the appropriate state agency are also often required, because most states have notification requirements similar to the federal law. Environmental managers should determine the specific state release requirements applicable to their facilities and remember that notification to the state agency does not satisfy the independent obligation to notify the federal National Response Center.

Federal and state agencies strictly enforce the requirement to ''immediately'' report specified hazardous substance releases. Penalties have been imposed by agencies against facilities that delay only one or two hours in making their release notifications. When in doubt, err on the side of prompt notification.

19.2 REMEDIAL ACTION

Superfund remedial actions typically involve the cleanup of contaminated soil or groundwater caused by the release of hazardous substances. Superfund establishes a broad scheme for imposing retroactive liability on classes of potentially responsible parties (PRPs) and creates a fund, known as the Hazardous Substance Superfund, to finance the cleanup of hazardous substance releases. Under Section 106 of Superfund, EPA is empowered to order PRPs to clean up or contain any release of hazardous substances or pollutants into the environment. EPA is also authorized to recover natural resource damages from responsible parties. Superfund identifies four types of parties potentially liable for cleanup costs:

- Current owners and operators of the contaminated facility, regardless of whether an owner or operator caused the release of the hazardous substance
- Any person who owned or operated the facility at the time of the disposal of the hazardous substance (i.e., past owners or operators)
- Any person who arranges for the treatment or disposal of the hazardous substance at a facility (i.e., generators)
- Any person who accepts or accepted any hazardous substance for transport to disposal facilities selected by that person (i.e., transporters)

Courts have uniformly held that liability under Superfund is retroactive, strict, and joint and several. Under the theory of retroactive liability, companies can be held liable for past actions that were legally acceptable at the time of the Act. Strict liability means the government need not prove any intent or negligence. The PRP is liable without regard to fault. Joint and several liability provides that each and every PRP is liable individually for the entire cost of cleanup. The only exception to joint and several liability is made when PRPs can convince a court that liability for cleanup costs can be equitably apportioned among the parties.

The Superfund law provides that PRPs can be held liable for a variety of response costs and damages. First, the federal government can recover all costs of removal or remedial actions so long as they are consistent with the National Contingency Plan (NCP). The NCP is the basic policy directive for federal response actions under Superfund and is published at 40 CFR Part 300 and subject to regular revision. If the PRPs fail to respond, the federal government can expend money from the Superfund to do the job. There is strong incentive for private parties to obey EPA's orders. If they refuse to clean up and later are found liable, the government can sue for reimbursement plus penalties and/or punitive damages up to triple the cost of cleanup.

PRPs can also be required to pay monetary damages for injury to, or loss of, natural resources resulting from releases of hazardous substances. Superfund provides for the appointment of federal and state "trustees," who have authority to assess natural resource damages and to bring actions against

PRPs in federal court to recover damages. In order to prevail, a trustee must prove that releases from the facility contributed to the natural resource loss for which the trustee is seeking recovery. Many experts believe that natural resource damage claims will form the basis of the next major round of Superfund litigation and may result in the payment of hundreds of millions of dollars in damages.

19.3 EPA'S EVALUATION OF SUPERFUND SITES

The major elements of EPA's system for evaluating Superfund cleanup actions are summarized as follows:

- *CERCLA Information System.* Upon notification or discovery of an actual or threatening hazardous substance release from a facility, EPA adds the facility to the CERCLA Information System (CERCLIS). CERCLIS is a computerized database containing the official inventory of potential and actual Superfund sites and is used by EPA to plan and track its response actions. CERCLIS consists of three inventories: (1) the Removal Inventory, (2) the Remedial Inventory, and (3) the Enforcement Inventory. Both active and inactive sites are included within the three inventories.
- *Hazard Ranking System.* If EPA decides that a site in CERCLIS requires additional action, EPA scores the site using the Hazard Ranking System (HRS). EPA developed the HRS in response to Superfund's requirement that the National Contingency Plan contain a system for determining response priorities among releases or threatened releases of hazardous substances. The HRS ranks a site by means of a mathematical rating scheme that combines the potential of the release to cause adverse environmental impacts with the severity/magnitude of these potential impacts and the number of people who may be affected.

  If a facility scores above a certain threshold, EPA will add the site to the National Priorities List (NPL). The NPL is a list of top priority hazardous waste sites eligible for cleanup with Superfund money. EPA ranks sites on the NPL based on their HRS scores and periodically revises the list to add new sites or delete old sites.
- *National Contingency Plan.* A key aspect of the EPA Superfund Program is the National Contingency Plan (NCP). The NCP contains EPA's regulations outlining the requirements for responding to actual or threatened releases of hazardous substances. The NCP describes methods of discovering and investigating hazardous substance releases, contains

guidance on how to carry out response action, and establishes site cleanup responsibilities. Any party undertaking cleanup at a site must comply with the NCP to preserve claims for cost recovery against other liable parties. The NCP is published at 40 CFR Part 300 and is subject to regular revision.

19.4 REMEDIAL INVESTIGATION/FEASIBILITY STUDY

Once a site has been selected for cleanup, the Remedial Investigation/Feasibility Study (RI/FS) process begins. The remedial investigation (RI) is the first part of the site remediation process. During an RI, basic information on the site is collected and potential remedial actions identified. The feasibility study (FS) uses this information to evaluate alternative remedial actions. The objective of the RI/FS process is to ensure that sufficient information is considered to support the evaluation of remedial alternatives and the selection of a remedy.

Following the completion of the RI/FS report, EPA prepares a Record of Decision (ROD) based on the RI/FS, any public comments, and other applicable agency guidance and regulations. The ROD documents the final remedy selection. The most common types of Superfund site remediation are on-site treatment and off-site disposal. On-site treatment technologies include soil vapor extraction, bioremediation, groundwater control and treatment systems, and on-site incineration. On-site remediation projects typically are not required to obtain state, federal, or local environmental permits. However, the remedy must comply with the substantive requirements that would have been included in such permits.

Operation and maintenance activities are initiated after the remedy is implemented. For privately funded cleanups, PRPs are responsible for operating and maintaining any treatment or containment system associated with the site for the life of the remedy. For Superfund-financed sites, a state must provide assurances that it will assume operation and maintenance responsibilities for the site. The Superfund law does not specify site cleanup standards but directs EPA to select cost-effective remedies that protect human health, welfare and the environment. As a general matter, Superfund response actions must attain or exceed Applicable or Relevant and Appropriate Requirements (ARARs) under federal environmental and public health laws. ARARs include cleanup standards, standards of control, and other environmental protection requirements. The ARARs process is a great source of confusion and debate relating to the evaluation and implementation of proposed remedial actions.

19.5 INVOLVEMENT WITH EPA AND PRP GROUPS

Companies generally become involved in a Superfund site in two ways. First, EPA may notify the PRPs of their potential liability and request the parties to undertake a response action or reimburse EPA for its costs incurred at a site. EPA's initial correspondence with PRPs is often a ''104(e) information request.'' Section 104(e) of Superfund grants EPA the authority to request and gather information in regard to a company's relationship to releases of hazardous substances. A 104(e) letter typically describes the site, sets forth a time period in which to respond, and contains a list of detailed questions regarding the site. Failure to make a timely response to 104(e) requests may subject a facility to an administrative compliance order or civil penalties. A facility's environmental manager who receives a 104(e) request should always consult with legal counsel before responding to the request, because 104(e) responses can be used as evidence in subsequent legal actions.

Once EPA has identified a group of PRPs, the agency typically issues either a general or a special notice letter regarding the site that contains a deadline for PRPs to respond. In cases where EPA determines that a site creates an ''imminent and substantial endangerment'' to public health and welfare and the environment, EPA is empowered to issue unilateral orders to PRPs requiring them to abate the danger or threat posed in order to protect human health and the environment. Finally, EPA may file ''cost recovery litigation'' to recover all of its costs and damages relating to the site.

Companies can also learn of their alleged involvement with a Superfund site through notification by a PRP Committee. PRP Committees initially consist of a small number of PRPs identified by EPA who are interested in bringing into the cleanup action as many other parties as possible to share the costs of cleanup. Although most of the negotiations with EPA and PRP Committees will be handled by the company's attorneys, environmental managers often play a critical role in determining the facility's historical involvement with a site and developing settlement strategies.

19.6 SETTLEMENTS UNDER SUPERFUND

PRPs must consider a number of factors in deciding whether to participate in a settlement of a Superfund case. Given Superfund's well-established retroactive, strict joint and several liability scheme, in most cases EPA can easily establish a PRP's liability. Unless the PRP can assert one of the limited defenses to Superfund liability, there is usually no good reason for the PRP to go to trial on the issue of legal liability. Consequently, most Superfund cases are resolved through out-of-court settlements.

Section 122 of Superfund contains EPA's negotiation and settlement provisions. Specifically, it authorizes EPA to enter into a variety of settlements with PRPs, including cash out, *de minimis*, and response action agreements. Section 122 also allows EPA to issue covenants not to sue, prepare nonbinding allocations of responsibility, and settle claims by arbitration. The company's legal counsel typically takes the lead role in negotiating Superfund settlements or litigating Superfund claims.

19.7 MINIMIZING SUPERFUND LIABILITY

The most likely involvement environmental managers will have with the Superfund Law is in minimizing a facility's exposure to Superfund liability. Because Superfund liability can be assessed against current owners of contaminated property, environmental managers should take measures to ensure that their companies do not unwittingly become owners of a contaminated piece of property. The 1986 SARA legislation added a new perspective to landowner liability under Superfund. Specifically, SARA provided that landowners who acquire property without knowing of any contamination at the site, and without any reason to know of such contamination, may have a defense to Superfund liability. This "innocent landowner" defense is subject to the following qualifications:

- *Need for Appropriate Inquiry.* At the time of acquisition, a landowner must have undertaken "all appropriate inquiry" into the previous ownership and uses of the property consistent with good commercial or customary practice in order to minimize liability. The statute enumerates a number of factors for the courts to consider in defining appropriate inquiry.
- *Continuing Disposal.* A party who acquires property upon which the disposal of hazardous substances continues after the acquisition will not be permitted to use the innocent landowner defense.
- *Subsequent Transfer.* A party will not be able to avail itself of the innocent landowner defense if it obtained actual knowledge of the release of hazardous substances and subsequently transferred ownership without disclosing this knowledge.

The innocent landowner provision in effect establishes a duty of inquiry, and standard of care to be exercised in such inquiry, in regard to environmental contamination in commercial real estate transactions. Since the enactment of the innocent landowner defense, environmental engineering consultants have established a variety of "environmental assessment" services to determine whether properties are contaminated by hazardous substances. A phased approach to environmental assessments is usually proposed by the consultant. A Phase I Assessment includes a visual inspection of the property, research of the seller's records, and examination of public records. It typically does not involve any environmental sampling. In the event that the Phase I Assessment indicates a possibility that hazardous substances are located on the property, the consultant usually recommends a Phase II Investigation. During the Phase II study, the consultant conducts environmental sampling to determine whether hazardous substances are present on the property.

The scope of the environmental inquiry will depend, of course, on the particular transaction and property involved. Although these investigations can be costly, the cost is usually justified in light of the potential legal exposure associated with environmentally contaminated properties. If significant environmental problems are discovered, parties may still want to proceed with the transaction by negotiating a purchase price that reflects the cost of cleaning up the property.

19.8 STATE SUPERFUND LEGISLATION

Most states have some form of state Superfund statute that creates an independent cleanup authority. Unlike many other pollution control laws, these cleanup statutes and legal authorities are not delegated or shared between EPA and the state government. It is therefore possible to face concurrent liabilities under both state and federal law or concurrent enforcement actions by both EPA and the state regulatory agency. As a result of this overlap, environmental managers must be familiar with both state and federal programs.

Enforcement authorities and liability schemes under state laws vary significantly. Most of the state liability schemes provide for retroactive, strict, and joint and several liability, just as the federal law does. The states commonly look to federal guidelines and standards to establish cleanup levels; however, some have enacted cleanup standards that greatly exceed the requirements of federal law. Further, some states, such as New Jersey, have enacted state laws that make closing a commercial real estate transaction contingent upon demonstrating to state environmental authorities that the property is not contaminated. The New Jersey state cleanup law, called the New Jersey Environmental Cleanup Responsibility Act (ECRA), is intended to ensure that contaminated property is discovered and cleaned up before it is transferred. Cleanup is made the responsibility of the transferring party and is a precondition to closure, sale, or transfer of the business. Several other states are considering ECRA-type laws.

20

EMERGENCY PLANNING AND COMMUNITY RIGHT-TO-KNOW

Statute:	42 USC §§ 11001 to 11050
Regulations:	40 CFR Parts 350 to 372

In 1984, public and congressional concerns regarding potential releases of hazardous substances from industrial sources and their impact on surrounding communities were heightened as the result of a major release of a toxic substance from a chemical plant in Bhopal, India. In 1986, Congress enacted a law to address these concerns as part of the Superfund Amendments and Reauthorization Act (SARA). Title III of SARA contains the Emergency Planning and Community Right-to-Know Act, commonly referred to as either the Community Right-to-Know Law or EPCRA. Under the provisions of EPCRA, the governor of each state was required to establish and provide for the operation of a State Emergency Response Commission (SERC). The commission, in turn, designated Local Emergency Planning Committees (LEPC), which typically included local emergency officials such as sheriffs and fire marshals, industry representatives, and other appropriate state or local officials. The local emergency planning committees devised emergency response plans for their areas, aimed at protecting the public and community in the event of an accidental release of regulated hazardous substances.

EPCRA established a variety of tools for the local emergency response committees to use. These tools include emergency reporting requirements, annual reporting of hazardous chemical inventories, and annual reporting of toxic chemical releases from facilities. The emergency release and annual reporting requirements have created significant responsibilities for the environmental manager at an industrial facility governed by the rules. When determining whether a facility is subject to the EPCRA requirements, managers should pay particular attention to the chemicals listed for each section of EPCRA, because listed substances for emergency reporting and annual release reporting differ. It also is possible for a facility to be subject to the emergency reporting requirements of EPCRA, but not the annual release reporting requirements, because of the particular chemicals and quantities of those chemicals used at the facility.

Since its inception, the Toxics Release Inventory (TRI) reports required by the Community Right-to-Know Law have created an annual scorecard that the public and the press use to compare releases of toxic chemicals from various industrial sources. It is common for newspapers to publish a ''top 10 list'' of facilities for each state, based on the information provided in these reports. Environmental managers with facilities on this list should anticipate questions and be prepared to discuss these reported releases with the public and the press.

EPA has also encouraged industry to voluntarily reduce releases of toxic chemicals. In the early 1990s, EPA's 33/50 program encouraged companies or individual facilities to voluntarily reduce releases of a list of 17 designated toxic chemicals. EPA's goal was to attain a 33% to 50% reduction in the releases of these chemicals over a period of years. Industry responded well to this initiative, and numerous companies successfully achieved the reductions proposed by EPA.

SARA has been amended several times since 1986. The chemical lists especially have been subject to change. Petitions to delist certain chemicals are constantly under review by EPA, and the agency has made significant additions to the original list of regulated substances. Facilities subject to the requirements of EPCRA should establish a process to monitor additions and/or deletions of listed chemicals.

20.1 EMERGENCY PLANNING AND NOTIFICATION

The emergency planning and notification requirements apply to any and all facilities where a listed extremely hazardous substance (EHS) is produced, used, or stored in quantities greater than the threshold planning quantity (TPQ), identified by the EHS list. Facilities must notify the SERC within 60 days and the local emergency planning committee within 30 days of becoming subject to the rules. The EHS list of chemicals and TPQs are identified in 40 CFR §§ 355.30(a) and (b). The entire EHS list contains more than 400 chemicals. Several listed chemicals and their associated TPQs are shown in Table 20.1.

Table 20.1 Examples of extremely hazardous substances

Chemical Name	CAS No.[a]	Reportable Quantity (lbs)	Threshold Planning Quantity (lbs)
Acrolein	107-02-08	1	500
Arsenic destolide	1303-28-2	1	100/10,000
Ethlythiocyanate	542-90-5	10,000	10,000
Hydrogen sulfide	7783-06-4	100	500
Phenol	108-95-2	1,000	500/10,000

[a]Chemical Abstract Service Number

Table 20.1 illustrates several important aspects of the emergency reporting requirements. The reportable quantity (RQ) may vary tremendously, from 1 pound to 10,000 pounds, depending on the substance. Likewise, the threshold planning quantities that are used to determine whether a facility is subject to EPCRA are likely to vary and are often different from the reportable quantities. In the case of solids, there are two TPQs. The lower number applies when a solid is handled in granular form (particles less than 100 microns), a solution, or molten form. The lower quantity also applies to solids with listed National Fire Prevention Act reactivity ratings of 2, 3, or 4. The higher TPQ applies to all other forms of the chemical.

20.1.1 Threshold Planning Quantity (TPQ) Determinations

The initial step in making TPQ determinations is to screen the EHS list and develop a potential list of chemicals. The TPQ for a given chemical is calculated on the basis of the total amount of the chemical present at the facility. If the quantity of the chemical present on-site could increase to a level that exceeds the TPQ at a later date, a procedure should be developed to identify this change. Because the facility becomes subject to the EPCRA requirements once it exceeds the thresholds, a review of mixtures of chemicals present at the facility is also required in determining the TPQ. In order to qualify for consideration in calculating the TPQ, a mixture or solid must have a concentration of 1% or more by weight of the chemical in question. Only the actual quantity of the chemical in the mixture or solid need be included in the TPQ calculation.

20.1.2 Notification Requirements

Facilities possessing extremely hazardous substances in quantities greater than the threshold planning quantity are required to notify the LEPC within

30 days. The facility must also designate an emergency response coordinator who will participate in the local emergency planning process.

Facilities that have EHS chemicals or substances above the TPQ are also required to report releases of these chemicals that exceed the RQ and are determined to present a potential risk of exposure to persons outside the facility property boundaries. The notification requirements do not apply to federally permitted releases or releases that are continuous and stable in quantity and rate. The RQ can be as low as one pound and the regulations require ''immediate'' notification. From a practical standpoint, the immediate notification requirement makes it very difficult to calculate accurately the amount of material released and whether the release resulted in potential exposure outside the property boundaries. Because of these uncertainties, the environmental manager is well-advised to err on the side of caution and make the necessary notifications in a timely manner with whatever information is available. As more data becomes available, the initial estimates provided to the agencies can be revised.

In addition to the EPCRA requirements, CERCLA also requires notifications for releases of ''hazardous substances'' to the environment. It is possible that a release will not trigger reporting obligations under Community Right-to-Know but will still require a notification to EPA's National Response Center under the requirements of CERCLA. For example, a release of CERCLA hazardous substance must be reported even if the release is totally contained within the facility boundaries. Accordingly, the facility should closely examine and compare the reporting requirements of both CERCLA and the Resource Conservation and Recovery Act (RCRA). The EPCRA notification requirements are generally summarized in Table 20.2.

The initial notification is to the LEPC and should include the chemical name, time, and duration of the release, an estimate of the amount of the release, known or anticipated health impacts, precautions taken as a result of the release, and names and phone numbers of facility contacts. The en-

Table 20.2 Release notification outline—EPCRA Section 304

A. Oral Notifications (immediate)
   1. Local emergency planning committee(s) and/or local emergency response personnel
   2. State Emergncy Response Commission
   3. National Response Center (1-800-424-8802) for CERCLA-qualified releases

B. Written Notifications
   1. Must be provided as soon as practicable after the event
   2. Include description of the release event, actions taken, known or anticipated health risks, and summary of medical treatment

vironmental manager or facility emergency coordinator should collect as much of this information as possible prior to making the notification.

20.2 CONTINUOUS RELEASES

An area of release reporting that became a concern to EPA after the enactment of EPCRA was how to account for releases of regulated chemical substances that occur continuously. For example, ammonia released through a tank vent as a normal part of a facility production process could exceed the 100-pound RQ on an ongoing basis. Both CERCLA and EPCRA provide notification procedures for an abnormal release of chemicals to the environment over a 24-hour period, but neither provides guidance on how to handle emissions that occur above the reportable quantity on a continuous basis. EPA eventually recognized this issue and in 1990 revised its regulations to allow for annual reporting of continuous releases of hazardous substances above the RQ. The continuous release report does not excuse a facility from reporting any upset or emergency release or spill of the substance beyond the normal operating conditions. In that event, the environmental manager would still be obligated to make the emergency release notifications described earlier in this chapter.

20.3 CHEMICAL INVENTORY REPORTS

The Community Right-to-Know Law also requires industrial facilities that are subject to the OSHA material safety data sheet (MSDS) requirements to submit information to the LEPC, the SERC, and the local fire department regarding hazardous chemicals stored on-site if the chemicals exceed certain threshold amounts. There are two steps involved in this process. First, under Section 311 of EPCRA, the facility is required to submit a list of hazardous chemicals subject to the MSDS requirement and stored in amounts exceeding 10,000 pounds during the preceding year. A newly introduced chemical should be added to the list within three months of the date it becomes subject to the requirement. If the chemical is included on the extremely hazardous substance list, the threshold is 500 pounds or the threshold planning quantity, whichever is less. Some states have established lower thresholds for all hazardous chemicals (e.g., Oregon, Louisiana). The facility environmental manager should verify the applicable state requirements prior to making a submittal.

The second inventory reporting requirement involves annual reporting of quantities of materials on-site and their location. These annual reports are due on March 1 of each calendar year and are based on the previous year's inventory. The report is to be submitted on EPA's Tier One or Tier Two forms. Although the Tier One form requires less information, in practice you will need to gather the same basic information to complete either form. The LEPC may request that facilities submit the more detailed Tier Two form. The form for Tier One reports is shown in Figure 20.1.

Page _____ of _____ pages

Revised June 1990

Form approved OMB No. 2050-0072

Tier One EMERGENCY AND HAZARDOUS CHEMICAL INVENTORY
Aggregate Information by Hazard Type

FOR OFFICIAL USE ONLY

ID#

Date Received

Important: Read instructions before completing form

Reporting Period From January 1 to December 31, 19___

Facility Identification

Name ______
Street ______
City ______ County ______ State ______ Zip ______

SIC Code [ ] Dun & Brad Number [ ]

Owner/Operator

Name ______
Mail. Address ______
Phone ______

Emergency Contacts

Name ______
Title ______
Phone ______
24 Hour Phone ______

Name ______
Title ______
Phone ______
24 Hour Phone ______

☐ Check if information below is identical to the information submitted last year.

	Hazard Type	*Max Amount*	*Average Daily Amount*	*Number of Days On Site*	*General Location* ☐ Check if site plan is attached
Physical Hazards	Fire				
	Sudden Release of Pressure				
	Reactivity				
Health Hazards	Immediate (acute)				
	Delayed (Chronic)				

Certification *(Read and sign after completing all sections)*

I certify under penalty of law that I have personally examined and am familiar with the information submitted in pages one through ____, and that based on my inquiry of those individuals responsible for obtaining the information, I believe that the submitted information is true, accurate and complete.

______
Name and official title of owner/operator OR owner/operator's authorized representative

______ Signature

______ Date signed

Report Ranges

Range Code	Weight Range in Pounds From....	To.....
01	0	99
02	100	999
03	1,000	9,999
04	10,000	99,999
05	100,000	999,999
06	1,000,000	9,999,999
07	10,000,000	49,999,999
08	50,000,000	99,999,999
09	100,000,000	499,999,999
10	500,000,000	999,999,999
11	1 billion	higher than 1 billion

254 Emergency Planning and Community Right-To-Know
\Chapter 20.doc 07/08/98

Figure 20.1 Tier One Form: Emergency and Hazardous Chemical Inventory

20.4 ANNUAL TOXIC CHEMICAL RELEASE REPORTS (FORM R)

Section 313 of the Community Right-to-Know Law created a program to address annual nonemergency releases of toxic chemicals into the environment. In general, facilities subject to the toxic chemical release reporting regulations are required to report annual releases of listed chemicals to the environment. The Section 313 requirements are based on yet another list of toxic chemicals, which differs from the lists discussed earlier in this chapter. Initially, the Section 313 list of toxic chemicals was based on the so-called Maryland/New Jersey list, but the list has had several revisions, including a major expansion of the list in 1995. The current toxic chemical list, published at 40 CFR § 372.65, should be monitored for changes on an ongoing basis inasmuch as new analytical techniques and the availability of test data results in frequent additions to the list.

20.4.1 Facilities Subject to Toxic Chemical Release Reporting

The applicability of the toxic chemical release reporting requirement is based on several qualifying criteria and, in general, is much narrower than other sections of EPCRA. Table 20.3 summarizes the applicability criteria.

Table 20.3 General applicability criteria/toxic chemical release reporting

The requirements of Section 313 of the EPCRA generally apply to facilities that meet the following criteria:

1. Facilities with 10 or more employees in the following Standard Industrial Classifications (SICs):
   - SIC Group 10—except 1011, 1081, and 1094
   - SIC Group 12—except 1241
   - SIC Groups 20–39
   - SIC 4911, 4931, 4939 (limited to facilities that combust coal and/or oil for power generation for distribution in commerce), 4953 (limited to facilities regulated under RCRA, Subtitle C)
   - SIC 5169, 5171
   - SIC 7389 (limited to facilities engaged in solvent recovery services on a contractual or fee basis)
2. Multiestablishment facilities (if greater than 50% of the value of products shipped and/or produced is from facilities within SIC Codes 20–39)
3. Facilities that manufacture, process, or otherwise use toxic chemicals in excess of the established threshold quantities

20.4.2 Threshold Determinations

If the environmental manager determines that the facility is subject to SARA 313 requirements, he or she must review the regulated substances and determine how they are used at the facility. This determination is based on whether the chemicals are "manufactured or processed" or "otherwise used." The threshold quantity is different for each category of use, as noted in Table 20.4.

The definitions of these categories are summarized as follows:

Otherwise used—Any use of a listed toxic chemical other than intentionally adding the chemical to a product distributed in commerce. Examples include chemical processing aids such as catalysts or solvents.

Manufactured—A listed toxic chemical that is produced, prepared, or imported. The definition of *manufactured* includes the coincidental production of toxic chemicals as by-products or impurities.

Processed—The preparation of a listed toxic chemical after its manufacture for distribution in commerce, regardless of changes in form or physical state, or whether it is part of an article.

EPA exempts some materials from consideration in determining whether the report thresholds have been exceeded. For example, *de minimis* levels of toxic chemicals in mixtures (below 1% concentration by weight or 0.1% for carcinogens), structural components, janitorial supplies, employee usage, motor vehicle maintenance, impurities from air and water intakes, and toxic chemicals contained in articles are not considered in calculating the threshold quantities. The actual calculations to determine the threshold quantities require extensive knowledge of the engineering and chemistry of the manufacturing process and annual facility production information. Typically, a team composed of representatives from environmental management, process engineering, purchasing, accounting, and operations is required to fully evaluate potential releases from a facility. Once the list of regulated chemicals is established for the facility, the various releases must be reviewed. Environmental managers should keep abreast of improvements to analytical techniques that may allow the facility to better measure its releases.

Table 20.4 Toxic release reporting threshold quantities

- Otherwise used chemicals—10,000 lb/year
- Manufactured or processed chemicals—25,000 lb/year

Table 20.5 EPA Form R report elements

1. General facility identification information, including relevant environmental permits
2. Identification of off-site disposal of toxic chemicals
3. Determination of whether the toxic chemical is manufactured, processed, or otherwise used
4. Estimate of the maximum amounts of the toxic chemical present at the facility
5. Annual quantity of the toxic chemical released to each environmental medium
6. Waste treatment or disposal methods
7. Pollution prevention actions (reporting is optional)

20.4.3 Form R Reports

Each chemical exceeding the threshold quantity must be analyzed to determine the releases to the environment that occurred during the calendar year. Reports are due annually on July 1 for releases that occurred during the previous calendar year. EPA has adjusted the date in some years because of modifications to the reporting forms and new reporting requirements, so it is important to verify the due date each year with the appropriate EPA regional office.

The toxic chemical release reporting form is commonly known as EPA Form R. Completing Form R reports also requires extensive knowledge of the manufacturing and production process to determine whether the release of a toxic chemical has occurred. Table 20.5 summarizes the elements of the Form R report. Form R itself is included in Appendix I.

A separate Form R is required for each chemical. Once complete, a senior official, normally the facility manager, must certify the accuracy and honesty of the report.

20.5 SUPPLIER NOTIFICATION REQUIREMENTS

The Community Right-to-Know Law requires facilities to notify customers if Section 313 toxic chemicals are contained in finished goods, mixtures, or trade name products. Supplier notifications are typically incorporated into the Material Safety Data Sheet (MSDS) prepared for products manufactured by the facility. If an MSDS is not required, the supplier notification must be provided on a separate form.

20.6 EPA'S TOXICS RELEASE INVENTORY

Each year EPA summarizes in a published report the 313 Form R reports submitted by the regulated industries. This report has become known as the

Top Ten Facilities for Total Releases

Facility	City, Parish	Air Emissions Pounds	Surface Discharges Pounds	Underground Injection Pounds	Releases Land Pounds	Total Releases Pounds
Cytec Ind. Inc.	Westwego, Jefferson	621,493	50,572	26,362,503	0	27,034,568
Arcadian Fertilizer L.P.	Geismar, Ascension	1,585,484	14,755,693	5	438,957	16,780,139
IMC-Agrico Co.	St. James, St. James	6,659,245	4,660,020	0	393,628	11,712,893
Rubicon Inc.	Geismar, Ascension	698,612	214	7,736,350	0	8,435,176
Monsanto Co.	Luling, St. Charles	66,321	234,050	5,397,660	0	5,698,031
CF Ind. Inc.	Donaldsonville, Ascension	4,958,095	610,605	0	0	5,568,700
Angus Chemical Co.	Sterlington, Ouachita	75,150	60,208	5,264,233	0	5,399,591
Cabot Corp.	Franklin, St. Mary	4,744,921	0	0	0	4,744,921
Cabot Corp.	Ville Platte, Evangeline	4,204,100	0	0	0	4,204,100
Witco Corp.	Harvey, Jefferson	18,100	0	3,870,000	0	3,888,100

Top Ten Facilities for Air/Water/Land Releases

Facility	City, Parish	Air Emissions Pounds	Surface Water Discharges Pounds	Releases to Land Pounds	Air/Water/ Land Releases Pounds
Arcadian Fertilizer L.P.	Geismar, Ascension	1,585,484	14,755,693	438,957	16,780,134
IMC-Agrico Co.	St. James, St. James	6,659,245	4,660,020	393,628	11,712,893
CF Ind. Inc.	Donaldsonville, Ascension	4,958,095	610,605	0	5,568,700
Cabot Corp.	Franklin, St. Mary	4,744,921	0	0	4,744,921
Cabot Corp.	Ville Platte, Evangeline	4,204,100	0	0	4,204,100
Triad Chemical	Donaldsonville, Ascension	3,298,182	35,161	0	3,333,343
International Paper	Mansfield, De Soto	3,190,411	336	710	3,191,457
Exxon Chemical	Baton Rouge, East Baton Rouge	1,965,933	987,909	0	2,953,842
BASF Corp.	Geismar, Ascension	740,925	2,207,718	0	2,948,643
Degussa Corp.	Louisa, St. Mary	2,570,000	0	0	2,570,000

FOR MORE INFORMATION:

State Contact: (504) 765-0737 — Linda Brown, Fax: (504) 765-0742
EPA Regional Contact: (214) 665-8013 — Warren Layne, Fax: (214) 665-7263

To obtain TRI data use assistance, call TRI User Support Service (TRI-US):
(202)260-1531 Fax: (202) 260-4659

Figure 20.2 U.S. EPA 1995 Toxics Release Inventory—Louisiana (Top Ten Facilities)

Population	4,338,072
Total Facilities	314
Total Forms	2,118
Form As	159
National Rank for Total Releases	2
National Rank for Air/Water/Land Releases	2
Transfers into State	
Rank	9
Pounds	54,597,771
Transfers out of State	
Rank	18
Pounds	38,943,904

Reported Releases and Waste Management Activities (pounds)	
On-Site Releases	
Air Emissions	84,841,485
Surface Water Discharges	28,268,576
Underground Injection	54,494,533
Releases to Land	4,654,598
On-Site Waste Management	1,863,490,788
Recycling	722,620,966
Energy Recovery	308,611,165
Treatment	832,258,657
Off-Site Transfers for Further Waste Management	77,724,509
Recycling	52,716,144
Energy Recovery	12,836,007
Treatment	9,840,800
Publicly Owned Treatment Works (POTWs)	44,015
Disposal	2,287,543
Other Off-Site Transfers	0

Top Five Chemicals for Total Releases

Chemical	Air Emissions Pounds	Surface Water Discharges Pounds	Underground Injection Pounds	Releases to Land Pounds	Total Releases Pounds
Ammonia	21,239,243	613,903	4,847,597	10,906	26,711,649
Methanol	16,065,357	340,255	5,660,169	231,688	22,297,469
Phosphoric acid	29,157	20,381,979	6,800	1,162,490	21,580,426
Nitrate compounds	0	6,338,585	6,758,000	12,519	13,109,104
Acetonitrile	24,526	583	11,003,500	0	11,028,609

Top Five Chemicals for Air/Water/Land Releases

Chemical	Air Emissions Pounds	Surface Water Discharges Pounds	Releases to Land Pounds	Total Releases Pounds
Ammonia	21,239,243	613,903	10,906	21,864,052
Phosphoric acid	29,157	20,381,979	1,162,490	21,573,626
Methanol	16,065,357	340,255	231,688	16,637,300
Carbon disulfide	7,884,116	0	0	7,884,116
Nitrate compounds	0	6,338,585	12,519	6,351,104

Figure 20.3 U.S. EPA 1995 Toxics Release Inventory—Louisiana (Reported Releases)

Toxics Release Inventory, or TRI report, and is available on the Internet. Once the data is summarized, it can be reviewed on an industry basis by state or by individual facility. Examples from the 1995 report are shown in Figures 20.2 and 20.3.

The annual TRI report is of great interest to industry and environmental groups because it presents an annual scorecard of toxic releases to the environment. Environmental activists often use the report to identify the top polluters in a given state or area of the country. In light of this use, it is advisable for facilities that release large quantities of toxic chemicals to prepare for possible questions from the press and the public about their releases. Although these releases may be permitted and determined to present no significant risk to human health or the environment, the overall amount of the release may still be of great concern to the public.

One word of caution regarding how to interpret the annual TRI report: Comparing total quantities of released chemicals can be deceiving. Several major factors influence the releases reported by industry. First, many more toxic substances are now included on the list than when the law first went into effect. Second, more knowledge and better analytical testing capabilities have allowed facilities to calculate releases that were "unknown" during the early years of the reporting program. On the other hand, many facilities have made process changes or modifications to control technology that have significantly reduced or eliminated annual releases. All of these factors should be considered in interpreting and comparing TRI data.

21

TOXIC SUBSTANCES CONTROL ACT (TSCA)

Statute:	15 USC § 2601 et seq.
Regulations:	40 CFR Parts 700 to 799

The Toxic Substances Control Act (TSCA) was enacted by Congress in 1976 in an effort to protect the public and the environment from an unreasonable risk of injury caused by chemicals. TSCA controls the manufacture, distribution, and sale of chemical substances and requires chemical manufacturers to understand and characterize the risks posed to human health and the environment by a chemical before that chemical is introduced into commerce. Under the authority of TSCA, the Environmental Protection Agency (EPA) has also promulgated detailed regulations regarding the use, management, storage, and disposal of polychlorinated biphenyl (PCB) materials. Because EPA administers the TSCA program, the environmental manager plays a key role at most facilities in managing TSCA compliance.

21.1 PREMANUFACTURE NOTIFICATION

TSCA requires manufacturers of new chemical substances to provide EPA with a Premanufacture Notification (PMN) 90 days before manufacturing activities begin. The PMN includes information about the chemical substance, how it is produced, categories of proposed use, and adverse health or environmental effects. Once EPA determines that the chemical will not present any unreasonable risks to human health or the environment, the chemical substance is placed on the TSCA Inventory and may be manufactured or imported by anyone. There are several detailed exceptions to the PMN procedures for research and development activities, exportation, and production of impurities, by-products, and other specified chemical substances.

Because the PMN requirements of TSCA relate only to those facilities that import or manufacture "new" chemical substances, the PMN regulations do not apply to the majority of American manufacturing facilities.

21.2 PROHIBITION OF USE

Section 15(2) of TSCA prohibits using for commercial purposes any chemical substance that a person knows or has reason to know was manufactured, processed, or distributed in violation of TSCA. Manufacturing facilities should develop a chemical purchasing procedure to ensure that the chemical substances it purchases are listed on the TSCA Inventory and authorized by TSCA. As a practical matter, TSCA inventory compliance efforts should focus on chemicals purchased from less than reputable manufacturers or new or experimental chemical formulations.

21.3 TESTING REQUIREMENTS

Section 4 of TSCA authorizes EPA to require testing of specific chemicals if EPA finds that (1) the chemical may present an "unreasonable" risk of injury to health or the environment, (2) there is insufficient data to determine the effects of the chemical on health and the environment, and (3) testing is necessary to develop the required data. The primary burden of this testing requirement falls on chemical manufacturers, and EPA has promulgated a comprehensive set of regulations describing when such testing will be required (see 40 CFR § 790.42).

21.4 REQUIRED RECORD KEEPING

EPA is granted broad authority under Section 8 of TSCA to require companies that manufacture, process, or distribute in commerce any chemical substance, to maintain a variety of records regarding the effects of those chemical substances. These record-keeping requirements are broader in scope than many of the provisions of TSCA, as they also apply to chemical ''processors,'' which are defined broadly in the statute to include any company that prepares a chemical substance or mixture for distribution in commerce. The three most important record-keeping obligations of Section 8 of TSCA are as follows:

- *Significant Adverse Reaction Reports.* Under the provisions of Section 8(c), companies that manufacture, process, or distribute in commerce any chemical substances or mixture must maintain records of the ''significant adverse reactions to health or the environment . . . alleged to have been caused by the substance or mixture.'' Certain manufacturing activities (e.g., mining of mineral ores and extraction of petroleum or natural gas) are exempt from these requirements. Consumer allegations of personal injury or harm to health, reports of occupational disease or injury, and reports of complaints of injury to the environment are all subject to this requirement. Records relating to the adverse reactions of employees must be retained for a period of 30 years. Any other adverse reaction record must be kept for 5 years. Adverse reaction records need not include commonly recognized health effects associated with a particular substance or those effects that can be directly attributable to an accidental spill. EPA has promulgated rules that outline the details of these record-keeping and reporting requirements. In general, companies are required to report significant adverse reaction information to EPA in response to a specific agency request.
- *Health and Safety Studies.* Under the provisions of Section 8(d), EPA has promulgated regulations directing chemical manufacturers, processors, or distributors to submit to EPA ''health and safety studies'' relating to selected chemical substances or mixtures. EPA has identified several hundred chemical substances for which companies must submit qualifying health and safety studies. The regulations identifying who is subject to this requirement and the listed chemical substances are found at 40 CFR Part 716.
- *Notification of Substantial Risk.* Section 8(e) of TSCA requires any chemical manufacturer, processor, or distributor who obtains information which reasonably supports the conclusion that such substance or mixture presents a ''substantial risk of injury to health or the environment' to immediately notify EPA of the information. This notification requirement does not apply if the company has actual knowledge that EPA is already aware of such information. Although EPA has not yet promulgated regulations detailing the requirements of Section 8(e), the agency expects companies to make these notifications within 15 days of the date the company receives the reportable information.

21.5 POLYCHLORINATED BIPHENYLS (PCBS)

PCBs are subject to a comprehensive regulatory program established by EPA under the authority of TSCA. Prior to the enactment of TSCA, PCBs were widely used as cooling and insulating fluids for electrical transformers, insulating medium for electrical capacitors, cooling fluids for electric motors, and hydraulic fluids. These fluids were generally referred to as ''askarels'' and known by commonly used trade names such as Pyranol, Aroclor, and Inerteen. Once EPA enacted its PCB regulatory program, American industry dramatically reduced its use of PCB materials, although PCB materials are still found in many older industrial facilities.

Under Section 6 (e) of TSCA, EPA is directed to promulgate regulations governing the manufacture, processing, distribution, disposal, storage, and marking of PCBs and PCB-containing items. Most of the provisions of the TSCA regulations governing PCBs are triggered only if the PCBs are in concentrations above a specified threshold level (generally 50 ppm). State programs often regulate PCBs at lower concentrations, and environmental managers should review their state regulatory programs to determine what, if any, authority the state has over PCB materials.

Under the federal regulatory program, when a transformer, capacitor, or any other electrical equipment has not been tested for its PCB concentration, EPA presumes a certain PCB concentration. For example, if a transformer does not have a nameplate or any information indicating the type of dielectric fluid in it, the transformer must be assumed to be PCB-contaminated equipment unless testing reveals otherwise. The requirements of EPA's PCB management program are briefly summarized as follows:

- *Use Restrictions.* TSCA provides that PCBs may be manufactured, processed, or distributed only in a ''totally enclosed manner'' that results in no exposure to humans or the environment. This general prohibition is subject to several important exceptions, which authorize certain uses of transformers and other electrical equipment subject to detailed restrictions and requirements relating to their management.
- *Marking and Inspection Requirements.* EPA regulations require that certain PCB equipment be marked according to specific regulatory requirements. PCB transformers that are in use or stored for reuse must be visually inspected on a quarterly basis for leaks of dielectric fluid. This inspection requirement does not apply to PCB-contaminated transformers (less than 500 ppm of PCB) or PCB capacitors. Records documenting these inspections must be maintained for at least three years after the disposal of the PCB transformer.

- *Operational Requirements.* All PCB transformers, including transformers in storage for reuse, must be registered with fire response personnel. The regulations also provide that no combustible materials may be stored within a PCB transformer enclosure and establish immediate notification requirements for any PCB material released during a fire-related incident. Finally, fluids leaking from PCB transformers must be properly cleaned within 48 hours and disposed of in accordance with regulatory requirements.
- *Storage and Disposal Requirements.* PCB materials must be disposed of within one year following the date they were placed in storage. PCB storage facilities must be constructed and sited in accordance with EPA regulatory requirements. EPA has developed a well-defined program for disposing of PCB waste materials. In general, when PCB materials are removed from service, they must be either incinerated, placed in a chemical landfill, or disposed of pursuant to some other approved alternate disposal method. Fluids with PCB concentrations greater than 50 ppm are typically incinerated.
- *Spill Cleanup Policy.* EPA's spill cleanup policy for PCB materials is found at 40 CFR § 761.12 et seq. This policy establishes strict guidelines to follow in the event of a PCB spill. Spills of PCBs are also subject to spill reporting requirements under the Comprehensive Environmental Response, Compensation, and Liability Act (CERCLA), the Clean Water Act, and Department of Transportation regulations. The EPA regional office typically serves as the lead agency in approving PCB spill cleanup plans.

21.5.1 PCB Management Programs

The liabilities and regulatory requirements associated with PCB management are so great that many facilities have decided to totally eliminate their use of PCB equipment. Companies typically develop schedules to systematically remove PCB equipment from service and replace it with non-PCB equipment over a period of years. Any PCB replacement program should pay strict attention to the storage and disposal requirements of the regulations.

22

SPILL PREVENTION AND RESPONSE PLANS

Statute (OPA 1990):	33 USC § 2701 et seq.
Regulation:	40 CFR Part 112

The storage of petroleum products on-site at an industrial facility is a common practice. Over the years, spill incidents have prompted the enactment of several laws and regulations that govern the storage of petroleum products. If a facility stores oil products on-site and is nearby a navigable waterway, a spill prevention plan will likely be required. Even industrial facilities that discharge into municipal treatment systems may be subject to local ordinances that require spill prevention measures to protect the integrity of the treatment system.

In response to a major diesel tank failure that resulted in a 750,000-gallon diesel fuel spill into Monogahela River in Pennsylvania, Congress passed the Oil Pollution Act of 1990 (OPA) and authorized the Environmental Protection Agency (EPA) to modify and improve its oil spill prevention program. Spill containment and response planning are the focal points of OPA and other federal and state regulations. The particulars of these programs are discussed in this chapter.

22.1 OIL SPILL PREVENTION PLAN APPLICABILITY

Section 311 of the Clean Water Act authorized EPA to require certain facilities to prepare and implement a plan to prevent the discharge of oil into navigable waters of the United States. This plan is referred to under the regulations as a Spill Prevention, Control, and Countermeasure (SPCC) Plan.

Owners or operators of industrial facilities are responsible for determining whether the SPCC requirements apply to their facility. A facility is potentially subject to the rule based on three factors: (1) it must be a nontransportation-related operation, (2) oil storage must exceed a specified quantity (detailed later in this section), and (3) the facility must reasonably be expected to create a spill situation that impacts a navigable waterway. In regard to the first factor, nontransportation facilities are defined by EPA as those listed in Table 22.1.

The next step in determining whether a facility is subject to the SPCC requirements is to conduct an inventory of the oil products stored on-site and determine whether the inventory exceeds the established thresholds. The environmental manager should add to the inventory the storage capacity and location of each oil storage tank or container at the facility, including oil stored in on-site pipelines, trucks, railcars, and other appurtenances, such as loading racks. The only exceptions from the inventory calculation are piping and hoses associated with handling or transferring oil products to and from a vessel. The inventory should be compiled on the basis of the facility's capacity to store oil products, not the actual quantity of material on hand.

Based on the completed inventory, the total amount of oil products stored must exceed the following thresholds to trigger SPCC requirements:

- Aboveground storage—Total storage capacity of 1,320 U.S. gallons or a single container in excess of 660 gallons
- Underground storage—Total storage capacity in excess of 42,000 U.S. gallons

The final step of the SPCC applicability analysis is to determine whether a spill could reasonably be expected to reach a navigable waterway. The def-

Table 22.1 List of nontransportation facilities

- Fixed onshore and offshore oil well drilling facilities
- Mobile onshore and offshore oil well drilling platforms, barges, trucks and other similar facilities (while in a fixed operating position)
- Mobile onshore and offshore oil production facilities (while in a fixed operating position)
- Oil refining facilities
- Oil storage facilities
- Industrial, commercial, agricultural, and public facilities
- Waste treatment facilities
- Loading racks, transfer hoses, loading arms, and appurtenant equipment
- Highway vehicles and railroad cars used to transport oil within the confines of a facility

inition of a navigable waterway is very broad. In practice, almost any waterway can be subject to the definition. The environmental manager should also note that the determining factor in whether a spill has the potential of reaching a waterway is solely geography. That is, the ability of man-made features such as dikes or other structures to contain a spill should not be considered in making this evaluation.

22.2 SPILL PLAN REQUIREMENTS

If you determine that your facility is subject to the rule, an SPCC plan must be prepared. Any existing regulated facility is immediately subject to the rule. In the case of a new or newly regulated facility, a plan must be prepared within six months and implemented within twelve months from the time of facility startup. EPA may authorize extensions of these deadlines. The SPCC plan must be kept on-site and made available to agency inspectors. It need not be sent to EPA for agency approval.

The emphasis of the SPCC plan is on spill prevention rather than spill cleanup and response. The plan should include procedures governing personnel training and supervision, maintenance, inspection schedules and procedures, and descriptions of the design and location of spill containment devices. Sample SPCC plans for regulated facilities are available from the American Petroleum Institute and other industry associations. EPA Region 10 has developed a sample SPCC plan based on a theoretical ''ABC Oil Company,'' which outlines the legal requirements for the plan. The sample plan is displayed in Figure 22.1.

22.3 SPCC PLAN CERTIFICATION AND APPROVAL

An SPCC plan must be certified by a registered engineer. The plan must also be signed and approved by a facility manager with authority to implement the plan. Once the plan is in effect, it is subject to review at least once every three years. Any required improvements should be incorporated in the plan within six months of the review. By way of example, a certification page prepared by EPA Region 10 is shown in Figure 22.2.

22.4 FACILITY SPILL RESPONSE PLANS

An onshore facility that receives or ships oil in bulk via marine vessels is subject to additional spill prevention and response regulations. The regulations establish a complicated set of criteria regarding which facilities are

ABC Oil Company
100 Neverspill
Post Office Box 100
Oilville, WA 98000
Telephone (123) 555-7890
Contact
John Doe, Owner and Manager

Certification:
Engineer:
Signature:
License Number: 0000-00 (Seal)
State: Washington
Date: 10 January 1974
1. Name and Ownership
   Name: ABC Oil Company
   100 Neverspill Road
   Post Office Box 100
   Oilville, WA 98000
   Telephone (123) 555-7890

Manager: John Doe
505 Oil Road
Oilville, WA 98000
Telephone: (123) 555-0987
Owner: Same
Other Personnel: Secretary-Bookkeeper
Dispatcher
Transport Driver
(3) Delivery People
Service Area: King County, Washington
2. Description of Facility
   The bulk of the ABC Oil Company handles, stores, and distributes petroleum products in the form of motor gasoline kerosene, and No. 2 fuel oil. The accompanying drawing shows the property boundaries and adjacent highway, drainage ditches, on-site buildings, and oil handling facilities.
Fixed Storage: (2) 20,000-gallon vertical tanks (premium gasoline)
(2) 20,000-gallon vertical tanks (regular gasoline)
(1) 20,000-gallon vertical tanks (kerosene)
Total: 140,000
Vehicles: (1) Transport Truck
(4) Tankwagon Delivery Trucks
The bulk plant is surrounded by steel security fencing, and the gate is locked when the plant is unattended. Two area lights are located in such positions as to illuminate the office and storage areas.
3. Past Spill Experiences: (none)
4. Spill Prevention—Storage Tanks

Figure 22.1 Sample SPCC Plan (Spill Prevention Control and Countermeasure Plan)

a. Each tank is UL-142 construction (aboveground use).

b. The main outlet valve on each tank is lock-shut when the plant is unattended.

c. Each tank is equipped with a direct-reading gauge.

d. Venting capacity is suitable for the fill and withdrawal rates.

e. Main power switch for pumps is located in a box that is locked when the bulk plant is unattended.

f. A dike surrounds the tank assembly. Its volume (height × area) is computed on the basis of a single, largest tank within (20,000 gallons), and allowance is made for all additional vertical tank displacement volumes below the dike height (estimated spill liquid level). Total storage capacity is 140,000 gallons. A two-inch water drain is located at the lowest point within the dike enclosure, and it connects to a normally closed gate valve outside the dike.

5. Spill Prevention—Vehicular

a. On-Site—The frontal highway ditch and the ditch on the property's southern boundary intersect before crossing the highway through a culvert, headed eastward and eventually to a stream located approximately one-half mile distant. Emergency containment action will constitute the erection of an earthen dam and placement of absorbent pillars at the entrance to the culvert. Additional cascading of barriers will be provided as necessary. Personnel training and drilling are described herein later.

b. Off-Site—Each vehicle is equipped with a shovel and two absorbent pillars. The driver is instructed to achieve emergency containment, if possible, then call the office for help immediately.

6. Personnel—All personnel have been instructed and rehearsed in the following spill prevention and countermeasure plans:

a. No tank compartments to be filled prior to checking reserves.

b. No pump operations unless attended continuously.

c. Warning signs are displayed to check for line disconnections before vehicle departures.

d. Instruction has been held on oil spill prevention, containment, and retrieval methods, and a "dry run" drill for an on-site vehicular spill incident has been conducted.

e. Instructions and phone numbers have been publicized and posted at the office regarding the report of a spill to the National Response Center (1-800-424-8802), the U.S. Coast Guard, the EPA, and the applicable state environmental agency.

f. Instructions and company regulations have been posted conspicuously, which relate to oil spill prevention and countermeasure procedures.

7. Future Spill Prevention Plans—By July 10, 1995 (implementation deadline), the following additional plans will be completed:

a. On-site storage of spill containment and retrieval materials and equipment: bagged absorbent, absorbent pillars, and booms. Storage facility will be well publicized and clearly identified.

b. Installation of a sand-filled catchment basin for minor routine spillage at loading pump intakes and at loading rack. Sand to be periodically replaced.

c. A routine inspection program with check-off listing of tanks, piping, valves, hoses, and pumps for the prevention of both major spills and minor spills or leakage through proper maintenance.

Figure 22.1 *(Continued)*

Certification Information

A. Name of Facility—Washington Bulk Storage Terminal

B. Type of Facility—Crude Oil Storage and Handling

C. Date of Initial Operation—10 January 1974

D. Location of Facility—1111 Main Street, Seattle, Washington

E. Name and Address of Owner—ABC Oil Company, P.O. Box 100, Oilville, Washington 98000

F. Designated Person Responsible for Oil Spill Prevention—John Doe

G. Oil Spill History—This facility has experienced no significant oil spill events during the 12 months prior to 10 January 1974.

H. Management Approval—Full approval is extended by management at a level with authority to commit the necessary resources to spill prevention.

Signature

Name: Ms. A. A. Jones

Title: President, ABC Oil Company

1. Certification—I hereby certify that I have examined the facility and, being familiar with the provisions of 40 CFR Part 112, attest that this SPCC Plan has been prepared in accordance with good engineering practices.

Signature:

Name: I. M. Tat

(Seal)

Date: 10 January 1974 Registration No.: 0000-00

State: Washington

Figure 22.2 Example certification page for an SPCC Plan

required to prepare a response plan. In general, facilities that conduct marine transfers of oil products and have a total oil storage capacity equal to or greater than 42,000 gallons, and facilities that have an oil storage capacity equal to or greater than 1 million gallons without adequate secondary spill containment, are required to prepare a spill response plan.

Facilities subject to the rule must determine the worst-case discharge volume and develop a spill plan that responds to the worst-case scenario. Extensive guidance is provided in the appendixes of the rule regarding how to determine the worst-case scenario. Spill response procedures are dependent on the number of oil storage tanks, type of oil products, and secondary containment design of the facility. In some areas of the country, associations or cooperatives have been formed to train and develop response teams, equipped with oil booms and oil recovery equipment, to contain and recover a spill. These cooperative efforts allow all of the regulated facilities within a given area to achieve compliance in a cost-effective manner.

Appendix J has been provided courtesy of Clean Sound, an oil spill cooperative that was created in 1971 to meet the spill response needs of industry in Puget Sound. The information given explains many of the practical aspects of preparing and implementing response plans for major oil spills.

23

ENVIRONMENTAL IMPACT STATEMENTS

Statute:	(NEPA) 42 USC §§ 4321 to 4370(d)
Regulations:	40 CFR Parts 1500 to 1508

On January 1, 1970, President Nixon signed into law the National Environmental Policy Act of 1969 (NEPA). NEPA is an extremely significant environmental statute in that it identifies environmental quality as a national priority and establishes procedures for incorporating environmental concerns into federal agency decision making. NEPA is unique in that it does not prohibit or regulate certain behavior. Rather, it requires federal agencies to follow procedures for evaluating the environmental impacts of a proposed project before approving the project. So long as the federal agency follows the procedural requirements of NEPA, there is no basis under NEPA for challenging the agency's ultimate decision. Most of the law governing NEPA has been provided by the courts through hundreds of judicial opinions, issued during the last 25 years, interpreting the procedural requirements of the law. A detailed discussion of those opinions is beyond the scope of this book. However, a general understanding of the NEPA process is important for any environmental manager because it is likely that the manager will become involved in an environmental impact analysis at some point in his or her career.

23.1 THE NEPA PROCESS

At the federal level, NEPA directs all agencies, to the fullest extent possible, to meet the following requirements:

- Utilize a systematic interdisciplinary approach in any agency planning and decision making that may affect the environment, and
- Include an Environmental Impact Statement (EIS) in every recommendation or report on proposals for legislation and other "major federal actions significantly affecting the quality of the human environment."

It is important to remember that NEPA does not apply directly to private entities. Rather, an EIS is required only for major federal actions that significantly affect the environment. Under the Council on Environmental Quality (CEQ) regulations, major federal action "includes actions with effects that may be major and which are potentially subject to federal control and responsibility." The term *major federal action* includes adoption of plans, regulations, and policy and approval of specific projects. The issuance of an environmental permit by a federal agency is an example of one activity that typically triggers the application of NEPA. If a project meets the "major federal action" requirement of NEPA, an EIS must be prepared unless the project qualifies for an exemption.

In response to the CEQ regulations, most federal agencies have adopted their own implementing procedures. These procedures generally provide that major federal actions are exempt from NEPA if the proposed project qualifies for a categorical exclusion or the agency prepared an Environmental Assessment (EA) that resulted in a "Finding of No Significant Impact." "Categorical exclusions" are categories of actions that the authorized federal agencies have predetermined will have no significant effect on the human environment. EAs are brief written reports prepared by federal agencies to determine whether to prepare an EIS or conclude that the project has no significant impact on the quality of the human environment. If the agency issues a Finding of No Significant Impact after the preparation of an adequate EA, the procedural requirements of NEPA have been fulfilled and the project may proceed. If the EA concludes that the project will have a significant impact, then the agency must prepare a complete EIS.

Table 23.1 Federal EIS process

- Define the action.
- Establish the documentation that is required.
- Formulate an approach to the EIS.
- Prepare the draft EIS.
- Obtain public review and comment.
- Prepare the final EIS.
- Prepare the Record of Decision.
- Implement the action.

If the preparation of an EI is necessary, the federal agency is required to follow the process outlir in Table 23.1.

23.2 REQUIRED C TENT OF AN EIS

NEPA requires th ronmental Impact Statements include the following elements:

- ntal impact of the proposed action''
- ''The e environmental effects which cannot be avoided should be implemented''
- ''Ar es to the proposed action''
- th ionship between local short-term uses of man's environment naintenance and enhancement of long-term productivity''
- irreversible and irretrievable commitment of resources which be involved in the proposed action should it be implemented''

Q regulations provide more detailed information regarding exactly evel of analysis must be included in an EIS.

e primary purpose of an EIS is to fully disclose and evaluate environmental consequences associated with the proposed agency action. Court decisions have interpreted NEPA to require that all possible environmental consequences of the proposed action be evaluated to allow agency decision makers to make an informed choice.

Agencies typically use the checklist approach to evaluate the environmental impacts of a proposed agency action. An example of an environmental impact assessment checklist is provided in Table 23.2.

An EIS must also analyze the benefits of the project in light of the environmental risks involved and compare the benefits and detriments of th proposed project with various alternative action scenarios. NEPA does n provide any guidance regarding what alternatives to the proposed action mu be considered by the agency.

Because NEPA requires only federal agencies to follow certain procedu and does not require the agencies to meet any established legal stand governing the merits of their decisions, courts may not rule on the wisd of the agency's decisions but may only evaluate whether the proced requirements of NEPA have been met.

23.3 PLANNING FOR NEPA COMPLIANCE

Once a proposed project is subject to NEPA review, compliance w requirements of NEPA can be a time-consuming process, which m

Table 23.2 Illustrative checklist approach to envi[illegible]mental impact assessment

	Construction Phase			*Operating Phase*		
Potential Impact Area	Adverse Effect	No Effect	Benefi- Effect	Adverse Effect	No Effect	Beneficial Effect
A. Land Transportation and Construction						
a. Compaction and settling						
b. Erosion						
c. Ground cover						
d. Deposition (sedimentation, precipitation)						
e. Stability (slides)						
f. Stress/strain (earthquake)						
g. Floods						
h. Waste control						
i. Drilling and blasting						
[illegible]. Operational failure						
and Use						
Open space						
Recreational						
gricultural						
sidential						
nmercial						
strial						
esources						
v						
on						
ater						
fur, carbon, ni-						
atter						
t						
t						
es						
rd						
om						
ral						
ems						

th the
y sig-

23.2 (*Continued*)

npact Area	*Construction Phase*			*Operating Phase*		
	Adverse Effect	No Effect	Beneficial Effect	Adverse Effect	No Effect	Beneficial Effect
ıtion Systems						
bile						
ation						

and D. Wooten, eds., *Environmental Impact Analysis.* © 1988 The McGraw-

s
st
of
on
cal
cant
ect.
s in-
view
state
on of
e after

nanager
nmental
ve to be
ency ap-

oval of the project. Consequently, one of the first
nalysis of a proposed company development or
ther NEPA review is required. If so, members
erstand the procedural requirements of NEPA
rocess.

ies that fail to comply carefully with the
face the prospect of lengthy and expensive
court injunction blocking the proposed
d expense can be avoided if the require-
n the outset and completely met. Al-
sion of an agency on its proposed ac-
roposed agency action because the
ith the procedural requirements of

23.4 STATE ENVIRONMENTAL POLICY ACTS

Several states have enacted state environmental review laws that are equivalent to NEPA, and it is at the state level that the environmental manager is most likely to confront an environmental impact analysis. California, New York, Minnesota, and Washington are four states with comprehensive state environmental policy acts. State environmental policy laws are usually structured similarly to NEPA but are often broader in scope. For example, the state of Washington has a State Environmental Policy Act (SEPA) that ha been in effect since 1971. Under the Washington SEPA, an agency must fir determine whether its proposed activity falls within the broad definition "major action." If so, SEPA's environmental review and documentati requirements apply unless the proposal fits within one of the "categori exemptions." If the proposal does not qualify for an exemption, the applic must prepare an "environmental checklist" describing the proposed pro As an illustrative example, the Washington Environmental Checklist i cluded as Appendix E (please note that these forms are under constant re and subject to future revision). Based on this checklist, the authorized agency decides whether the proposal is significant. If a Determinat Significance is issued, project approval can be issued only by the stat the preparation of a draft and final Environmental Impact Statement.

For purposes of project permit planning, the environmental m should determine whether the state in question has a state enviro policy statute. If so, it is quite likely that the EIS process will ha followed before any significant state environmental permits or ap provals are issued.

24

RISK ASSESSMENTS

Rising public concern regarding the potential for adverse impacts on human health and ecological systems caused by environmental pollution has driven agencies to look for new methods to judge the hazards associated with environmental emissions and discharges. Environmental agencies are frequently using risk-based analysis in assessing these environmental impacts. Risk assessments usually focus on the level of risk posed to human health and the environment by exposure to a chemical substance over time. Risk assessments typically take a multimedia approach, adding the exposure risks resulting from air emissions and water discharges of a particular substance. They also can be multi-pathway assessments, evaluating exposures resulting from breathing in a substance directly from the air and ingesting the substance through the food chain.

Risk assessments are focused on particular chemical substances of concern. These are typically substances that have been listed by federal or state agencies as carcinogenic, mutagenic, or hazardous to human health for some other reason. In the case of air toxics, many states have established "fence line" or property line screening levels to act as a trigger for further evaluation. A property line screening level is a level of air toxic emissions that is calculated to protect the health of persons living off the plant site. If a toxic or hazardous air pollutant exceeds the established screening level, a full risk assessment may be needed to demonstrate whether the activity falls within an acceptable level of risk. Determining what constitutes what an acceptable risk can be very subjective. In the case of a carcinogen, typical acceptable risk levels may be noted as "1 in 100,000" or "1 in 1,000,000," meaning up to one additional cancer occurrence in a population of that many exposures.

Noncarcinogenic hazardous pollutants are normally judged by an acceptable index comparison or concentration requirement.

Environmental managers typically become involved in environmental risk assessments under the following circumstances:

- *Project Permitting*. Environmental agencies may require a risk assessment of one or more toxic pollutants emitted, discharged, and/or disposed of by the proposed facility.
- *Remediation Studies*. Risk assessments are commonly used to determine cleanup levels for contaminated soil.
- *Ecological Studies*. An agency may require a study of the ecological impacts to animals and fish resulting from the bioaccumulation of a discharged pollutant or local concern for an endangered species.

The risk assessment process is a very complex science. Most facilities do not have in-house toxicologists or biologists who are familiar with the scientific protocols and can negotiate the requirements of risk assessment with the regulatory agency. The success of a risk model is based on using the assumptions or exposure scenarios that are appropriate for a specific facility. It is strongly recommended that any facility dealing with a significant risk assessment question retain an expert who is very familiar with risk assessment methodology and who can work with the regulatory agency in developing the initial protocol and interpreting the results of the assessment.

24.1 RISK ASSESSMENT METHODOLOGY

As the use of risk assessments by regulatory agencies grows, some standardization of the methodology has been established. The National Research Council issued risk assessment recommendations in the early 1980s. Driven by the requirements of Superfund, the Environmental Protection Agency (EPA) also published an exposure assessment manual and continues to issue risk assessment guidelines. Based on guidance provided by the National Research Council and EPA, a human health risk assessment contains four basic elements:

- Hazard identification
- Exposure assessment
- Toxicity assessment
- Risk characterization

Although the general organization of the risk assessment process may be standard, many of the specific requirements are not. Prior to beginning work on any risk assessment, the environmental manager should meet with the agency staff involved in reviewing the assessment and reach agreement as to what chemicals, pathways of exposure, and other factors will be included in the risk assessment.

24.1.1 Hazard Identification

The initial step in the hazard identification process is to define the substances of concern to the agency. The applicable federal and state regulations for toxic pollutants provide the basis for this review. Other substances can be listed on the basis of site-specific concerns or available scientific literature relating to the manufacturing process used at the site.

Next, the exposed populations should be identified. The determination of exposed populations is often an iterative process. For example, if air dispersion modeling has not been completed, the determination of what areas should be considered in the risk assessment will be only an estimate. Later in the evaluation, the areas may be adjusted on the basis of reductions in estimated emissions owing to process changes or added environmental controls.

24.1.2 Exposure Assessment

Identification of human exposure pathways is the next step in the health risk assessment process. The substance and method of release (air, water, etc.) are the major factors in the pathway selection. Table 24.1 lists some of the potential exposure pathways considered in a typical risk assessment.

Table 24.1 Typical exposure pathways

DIRECT PATHWAYS

- Dermal contact with contaminated soil
- Dermal contact with contaminated surface water
- Inhalation of airborne contaminants
- Ingestion of contaminated soil (infants and children)
- Ingestion of contaminated drinking water
- Ingestion of contaminated surface water

INDIRECT PATHWAYS

- Ingestion of contaminated fish
- Ingestion of contaminated produce
- Ingestion of contaminated mother's milk

Each of the relevant exposure pathways for the risk assessment will have to be evaluated for each chemical or substance of concern. This evaluation involves determining the concentration of the substance as it travels from the source through the use of maximum airborne concentrations, soil deposition concentrations, or water dispersion models. These sophisticated models will provide the data to estimate the intake exposure required for the development of the risk estimates.

24.1.3 Toxicity Assessment

A toxicity assessment entails a review of the available scientific information regarding the toxicity of each substance under study. For example, scientific data will indicate the number of tumors or birth defects observed at various ingestion levels in a test animal. Predictions of carcinogenity or toxicity can also be made, based on a review of this data. This toxicity information supports the risk assessment by providing the quantitative input to the dose-response model used by the toxicologist to assess the range of risk.

Toxicity study data is not available for all substances. The quality of an available study or the conclusions reached in the study may also be subject to question. For example, the carcinogenic potency for substances like chloroform and dioxin have been debated by scientists for many years. Before an investigator agrees to evaluate the risk of a substance, a preliminary assessment of the available toxicity data should be conducted. Any disagreement regarding predicted toxicity should be resolved with the regulatory agency early in the risk assessment process.

24.1.4 Risk Characterization

Finally, the risk assessment must estimate the effects to human health of the substances under study. This will include increases in carcinogenic risk levels for substances listed as carcinogens and the potential chronic health effects for those substances listed as noncarcinogens. The carcinogenic risk, by design, is for the most exposed individual over a lifetime (standardized at 70 years), which tends to be a conservative evaluation. The noncarcinogenic risk level is developed by comparison of exposure levels to those levels identified as permissible by the local or state health department. An example of a risk characterization summary for airborne inhalation is shown in Table 24.2. As you will note from the table, acetone does not appear to present a major health hazard. Chloroform emissions, on the other hand, add 0.9 potential lifetime cancers in a population of 100,000.

Ideally, the risk assessment will indicate that all chemicals or substances under study are of such low concentrations that exposure results in no adverse

Table 24.2 Example risk assessment summary

	Inhalation of Air Contaminants			
Emitted Pollutants	Predicted Daily Intake (mg/kg/day)	Acceptable Chronic Level (mg/kg/day)	Hazard Index	Excess Lifetime Cancer Risk
Acetone	0.0009	3.00	0.0003	[a]
Chloroform	0.000075	[b]	—	0.9×10^{-5}

[a] Acetone is not considered carcinogenic.
[b] Chloroform is a probable carcinogen.

impacts. In the case of carcinogens, most states have concluded that an increased risk of cancer is acceptable if the risk is less than 1 in 1,000,000, often displayed in scientific notation as 1×10^{-6}. Most states have procedures for allowing greater levels of risk under certain circumstances.

24.2 ECOLOGICAL EXPOSURE ASSESSMENT

The risk assessment procedure can also include ecological exposure assessments when there are concerns relating to possible adverse impacts to neighboring wildlife and their habitat. Environmental activists have increasingly focused on ecological impacts in an effort to decrease releases of toxic materials. Environmental agencies are requiring more ecological exposure assessments as more species are listed as endangered or threatened under the Endangered Species Act. Ecological risk assessments identify key indicator species of concern, review the exposure pathways, and quantify the exposures. Exposure pathways are likely to include food chain events. For example, a toxic substance in a facility's discharge may bioaccumulate in the flesh or lipid of a fish. These fish, in turn, are eaten by a species of bird that further assimilates the toxic substance. The final ecological impact may result in a direct health impact to the adult bird or possibly manifest itself as a detrimental effect in the bird's eggs and/or reproductive cycle. The pesticide DDT was banned for its adverse impact on bird reproduction as a result of the thinning of eggshells in bird populations exposed over time to the chemical.

24.2.1 Species Selection

The selection of an indicator species selected for an ecological risk assessment is likely to be a very site-specific determination. Species selection proposals are usually made by a facility to the environmental agency for its review and approval. Environmental activist groups are also likely to play a role in the selection process if the evaluation relates to a controversial project or facility. Before a facility agrees to the selection of the indicator species, it should confirm that there is enough scientific data available to conduct a study. The following are criteria that may be useful in the selection process:

- Data availability for bioaccumulation, ecotoxicological concentrations, etc., in published studies or through environmental agency sources
- Bioaccumulation potential of the species for the substance of concern
- Likelihood of the species reliably representing local conditions

24.2.2 Exposure Concentrations

In making a determination of potential harm to a selected indicator species, the ultimate goal is to develop a concentration estimate of the substance of concern in the body of the species under study. This concentration will be compared with a body concentration amount known as the ecotoxicological benchmark concentration. The ecotoxicological benchmark concentration comes from an independent scientific study and indicates the concentration of the substance at which scientists observed harmful effects to the species being studied. For example, in fish studies, reproductive problems or mutagenic impacts to minnows may increase above the benchmark concentration; in bird studies, hatch rates may be adversely affected by the thinning of eggshells.

The following example is illustrative of the process. Assume that the indicator species selected for study is a bird that lives in the vicinity of a river into which a proposed industrial facility will discharge its effluent. This species of bird is known to eat fish from the river. From knowledge of the facility's process and the effluent treatment system, the environmental manager has determined the level of discharge of each chemical substance. The next step is to determine the level of bioaccumulation of the chemical of concern in the fish. Bioaccumulation is the extent to which chemicals accumulate in living organisms such as fish and other animals. The level accumulating in the fish can be determined by scientific models of the concentration of the chemical discharge and knowledge of the bioaccumulation factor for the species of fish under study. This analysis enables the risk assessment team to calculate concentration of the chemical in the fish's body. Once the level of contaminant in the fish is established, further calculations can then be made by biologists estimating the body concentration in the bird. These calculations are based on knowledge of how many fish are eaten by the bird in a given period of time. The concentration calculated in the bird's body as a result of fish consumption is then compared with the ecotoxicological benchmark concentration. If the ratio is higher than 1.0, modifications to the facility will likely be required to reduce the discharge of the chemical substance under review.

As in human health risk studies, it is important for the environmental manager to determine whether the necessary scientific information is available prior to agreeing to conduct an ecological assessment. If no knowledge exists concerning the benchmark concentration or bioaccumulation factor for a given species, it would be impossible to conduct an ecological risk assessment in a timely and cost-effective manner.

24.3 UNCERTAINTIES

The risk assessment process normally includes a discussion of uncertainties. The summary of uncertainties will review the potential for error in the evaluation and examine the quality of data and information, the uncertainty in the facility's emission or discharge estimates, and whatever other assumptions were used in the risk assessment. The objective of this exercise is to thoroughly evaluate the completeness and accuracy of the risk assessment. For instance, if extrapolations from scientific studies of other species were used, this is the section of the report where they should be noted. Another example may be the extrapolation of dosage data outside the experimental range; is the resulting effect linear? If the assessment assumed it was, but the real answer is unknown, this too should be noted in the "uncertainties" section of the risk assessment.

24.4 RISK ASSESSMENT COMMUNICATIONS

Normally, risk assessments are associated with a public process such as an application to permit a new facility or renew an existing permit. The environmental manager should consult and involve the facility manager and communications manager in this process and determine how best to present the results of the risk assessment to employees and the general public. The public is extremely interested in human health impacts and the health of endangered wildlife. Unfortunately, it is often difficult to communicate that risk levels for a facility are acceptable, because any increase in risk can be viewed negatively by the public. The public also draws distinctions between "involuntary" and "voluntary" risks. A person can choose to drive on a crowded freeway or fly in a plane, but that person does not have a choice in accepting the risks associated with the operation of a new industrial facility. In communicating risks, it is helpful to compare the risks of a project with the level of risk established for other common events. Some common lifestyle events, such as driving a car or drinking a beer, often provide a much higher risk over time than the environmental releases from a new facility. To achieve greatest acceptance, communication experts advise companies to communicate the results of risk assessments to the public in a simple and straightforward manner.

25

BASIC ENVIRONMENTAL MANAGEMENT PROGRAMS

Sound environmental management programs should both ensure regulatory compliance and adhere to certain management fundamentals. Most agency inspections focus on reporting requirements, records, and permit exceedances and do not emphasize actual pollution control, pollution prevention, or innovative management techniques. In this chapter we outline the basic elements of an environmental management program that will allow you to satisfy regulatory agencies, plant management, and the general public.

25.1 DATA COLLECTION AND RECORD KEEPING

The environmental permits issued to your facility are the starting point for your data and record-keeping requirements. Make a detailed list of each permit requirement, note whether the data must be reported to the regulatory agency or only kept on file, and determine how often the data must be collected and reported. Update this master list every time a permit is modified or a new permit is issued. In addition to your environmental permits, you also should review other environmental laws and regulations to determine whether there are additional record-keeping requirements. There are many requirements that will not necessarily be detailed in your permits. The following are examples of these requirements:

- Toxic Substances Control Act (TSCA)—Records for polychlorinated biphenyl (PCB) equipment removal must be retained for five years from the date of removal.

- Clean Air Act (CAA)—Asbestos removal and disposal records and agency demolition notifications should be retained.
- Resource Conservation and Recovery Act (RCRA)—Manifests of hazardous waste shipments must be retained for a period of three years. Waste characterization determinations must also be kept on file.
- Emergency Planning and Community Right-to-Know Act (EPCRA)—Calculations that support the conclusion that chemical use or releases are below the reporting thresholds should be retained at the facility.

For records that are kept in the field, such as via recorders or logs, ensure that periodic field checks are conducted. For example, if a road sweeper requires a driver logbook to verify fugitive dust control in your air operating permit, does the driver make entries on a continuous basis? When the logbook is full, does the operator know where to take it for filing? Are the field recorders inspected periodically, and is there a system to ensure data is taken to the file?

In today's world of computer networks and process information systems, automated data collection systems are evolving quickly. Always ensure that there is a backup to your data system and that data and records can be reproduced in the event of a system malfunction. Inexpensive tape or "zip" drives are among the tools that can be used to create backups periodically.

A good records management system will include permits, monitoring reports, administrative consent orders, and any other documents essential to managing the facility's environmental program. Environmental documents typically include the following:

- Copies of all pertinent statutes and regulations (federal, state, and local)
- Copies of all operating permits and consent orders
- Copies of all monitoring reports and inspection activities
- Copies of all relevant correspondence between the facility and federal, state, and local regulatory agencies
- Copies of emergency response plans and other critical facility documents
- Copies of environmental claims and complaints and the facility's responses

It also is important for the facility to establish an appropriate retention schedule for its environmental documents and strictly adhere to it. Technical staff are known for keeping records and data beyond their useful life, so care should be taken to ensure that these personal working files are subject to the same document control and retention policies. If exceptions are made to the retention policy, the value of the program diminishes significantly.

There is no regulatory requirement regarding how to establish an environmental records filing system. Most well-managed facilities centralize their environmental files. Scattered responsibility for management of environmental records increases the risk of errors, voids, or misplaced material. Appendix C provides a sample filing index and records retention schedule for an industrial facility.

25.2 MONITORING AND ANALYTICAL TESTING

Environmental compliance data is often gathered by monitoring instruments or analytical testing. Environmental managers will have to determine whether their facilities are capable of performing these activities in a cost-effective manner. Industrial facilities frequently use a combination of in-house and contracted services to meet their monitoring and analytical testing needs. Monitoring requirements are either continuous or periodic in nature. Typical monitoring requirements in each category are as follows:

Continuous Monitoring

- Air emission monitors for such pollutants as opacity, total reduced sulfur (TRS), or sulfur dioxide (SO_2)
- Measurement of pH, flow, or conductivity in water discharges
- Data relating to process parameters such as pressure or temperature

Periodic Monitoring

- Annual air emissions tests
- Quarterly groundwater well samples
- National Pollutant Discharge Elimination System (NPDES) permit renewal priority pollutant testing

If the facility operates continuous monitoring equipment, it is essential to have immediate access to skilled instrument technicians to allow for recalibration or repairs if the equipment malfunctions. In addition, the environmental manager will have to schedule periodic recalibrations and testing of these instruments as required by the applicable permits.

Periodic analytical testing will also be required at most facilities. If the facility has a chemical laboratory, some basic analytical testing can be conducted on-site. If the tests are complicated or require special test equipment, most environmental managers send their samples to certified commercial labs. Not all commercial labs produce the same quality of test results, and quality control in any lab can vary as personnel changes occur. Accordingly, the environmental manager should periodically qualify any off site testing labs and ask the labs to supply their detailed QA/QC procedures. Examples of tests sent to off-site laboratories include effluent toxicity tests conducted with fish or marine animals and chemical contaminant testing requiring high-resolution mass spectrometers. In the pulp and paper industry, for instance, dioxin testing in mill effluents requires detection levels in the low parts per quadrillion range. This type of testing can be done only by very sophisticated labs.

The Environmental Protection Agency (EPA) conducts research on test methods and certifies acceptable testing procedures. These methods are often referred to by an EPA method number. For example, EPA's protocol for testing air particulate matter is known as Method 5. EPA has established hundreds of different test methods that are either referenced in the pertinent regulations or published on EPA's website in the Index to EPA Test Methods. Industry technical associations often work closely and cooperatively with EPA on test procedures to help ensure the practicality and accuracy of testing methods.

25.3 REPORTING

The environmental manager's obligations to report all incidents, releases, and spills in a timely and proper manner have increased significantly owing to expanded legal requirements and higher public expectations. Proper reporting of spills and other environmental releases has become a high priority within federal and state environmental regulatory agencies. Spill reporting requirements are found in a multitude of federal and state laws, including the federal Clean Water Act (CWA), the Resource Conservation and Recovery Act, the federal Superfund Law, the Emergency Planning and Community Right-to-Know Act, and the Toxic Substances Control Act.

To ensure proper reporting, a facility should first identify all the legal requirements that apply to the facility. When these reporting obligations have been established, the facility should then develop its own procedures for reporting spills to regulatory agencies, employees, and the general public. All communications regarding an emergency spill or release should be clear, appropriate, and accurate. Once a spill reporting plan is established, it should be frequently reviewed and revised because out-of-date plans are potential legal violations and sure embarrassments in the event of an actual spill or release event.

When in doubt as to whether to notify a regulatory agency, err on the side of making the notification. In situations where a formal report is not required by law, consider informing the relevant regulatory agencies of the release informally. Such notifications build credibility with the agencies and prepare them for the public questions that invariably accompany any significant spill or release event.

Environmental permits, statutes, and regulations require periodic reporting of air emissions, water discharges, and other routine environmental releases. It is important for facilities to develop a management program to ensure that complete records are submitted to the appropriate regulatory agency in a timely manner. A basic outline of chemical or regulated substance reporting requirements is shown in Table 25.1. Reporting spills or releases is a full-time, 7-day, 24-hour responsibility, and the facility must develop a foolproof system to meet its notification obligations.

The proper reporting of regulated chemical substance spills or releases is complicated by the lack of consistent reporting obligations throughout the regulations. As a result of this confusion, many facilities have developed their own site-specific list for reporting. An example of such a list is shown in Table 25.2. Facilities that prepare their own listings should also develop a program to periodically review and revise the list.

Table 25.1 Emergency spill and release reporting requirements[1]

Statute/Regulation	Local Emergency Coordinator	Report to State or Local Agency[a]	National Response Center[b]
CAA		X[o]	
CERCLA[d]			X
CWA		X[c]	X
EPCRA	X	X[e]	
OPA[f]	X	X[e]	X
RCRA			X[g]
RMP[h]	X	X[e]	

[a] Verify your specific state, region or local agency contacts.
[b] Phone: 1-800-424-8802.
[c] Permits or local rules may add further reporting requirements.
[d] Comprehensive Environmental Response, Compensation, and Liability Act.
[e] State Emergency Response Commission.
[f] Oil Pollution Act
[g] Or phone On-Scene Coordinator designated in the local contingency plan.
[h] Risk Management Plan

Table 25.2 Example chemical reportable quantity (RQ) List (lb)

Chemical Name	SARA 302[a]	CERCLA	CAA
	RQ	RQ	TQ
Acetaldehyde	—	1,000	10,000
Acetone	—	5,000	—
Acronein	—	1	5,000
Benzyl cyanide	500	—	—
Bromine	500	—	10,000
Chlorine	—	10	2,500
Cyclohexanamine	10,000	—	15,000
Ethyl cyanide	—	10	10,000
Fluorine	—	10	1,000
Formaldehyde	—	100	15,000
Hydrogen peroxide (>52%)	1,000	—	—
Methane, chloro	—	100	10,000
PCBs	—	1	—
Sodium hydroxide	—	1,000	—
Toluene	—	1,000	—
Vanadium pentoxide	—	1,000	—
Zinc phosphide	—	100	—

[a] Superfund Amendments and Reauthorization Act of 1986

25.4 SCHEDULES AND PLANNING

A comprehensive environmental calendar and reminder system should be established for your facility. The environmental calendar should include all agency reporting requirements, training schedules, equipment calibrations, permit renewal dates, capital planning review dates, lab certification requirements, and any other relevant internal or external deadline activities. The calendar should also include a ''tickler,'' or advance warning system, to remind you of compliance deadlines and provide time to prepare reports sufficiently ahead of the required submittal dates to allow time for internal review. Failing to submit required reports and missing reporting deadlines are both high-visibility issues for environmental inspectors.

An environmental calendar is effective only if it is updated continuously. Environmental staff members should be required to periodically review the calendar, and a procedure should be established to ensure that any new requirements are added to the calendar. Once established, the environmental calendar can be used by all facility environmental managers to ensure that key compliance dates and obligations are never missed.

25.5 MANAGING PERMIT COMPLIANCE

A major function of the environmental manager is to ensure compliance with the provisions of the environmental permits. This responsibility typically involves maintaining compliance with existing permits, renewing existing permits in a timely manner, and either obtaining new permits or modifying existing permits. Maintaining permit compliance is the primary focus of most environmental management programs and is discussed in detail throughout this book.

Renewal applications are required for a variety of permits. Examples of renewal obligations are NPDES water discharge permits (renewed every five years), Title V air operating permits (renewed every five years), and solid waste landfill permits (typically renewed annually or up to every five years). Because environmental permit renewal activities usually begin months before an existing permit expires, the environmental calendar permit renewal dates should include enough time to complete these activities. For example, NPDES permits require that the renewal application be submitted six months in advance of expiration. The application requires that a series of ''priority'' pollutants be tested and those test results included as part of the application. Consequently, the environmental manager must begin work on the renewal process almost a year in advance of the permit expiration date. If proposed changes in the existing discharge limits are part of the renewal process, this too will increase the amount of time required to obtain the permit.

The environmental manager must also develop a system for obtaining permits for new or modified projects at the facility. Permitting new projects often requires changes to several permits and significant commitments of time and resources. The processes and requirements for obtaining new permits for air, water, solid waste, and other environmental media are discussed in detail in the appropriate sections of this book.

25.6 PROCESS KNOWLEDGE

As the focus of pollution control has shifted from end-of-pipe technology and treatment to pollution prevention at the source, it has become essential for the environmental manager to develop a detailed understanding of the facility's manufacturing process and process chemistry. Knowledge of the various processes is also essential in establishing credibility with the facility's operating managers and hourly employees. Production and business planning experience are important components of any environmental manager's career development plan.

Technical knowledge of a process must be supplemented with an understanding of the basic economics of the facility's business. The environmental manager must determine how best to achieve environmental compliance within the economic realities of the facility, and a clear understanding of these realities is critical to the success of the environmental management program. Knowing the business will also allow the environmental manager to put appropriation requests, control technology costs, and the environmental department budget into a broader context.

25.7 CAPITAL PROGRAM INPUT

The usual trigger for obtaining new or modified environmental permits is a process change or capital project. In most companies the best way for the environmental manager to become aware of those changes in a timely manner is to be part of the strategic planning and capital budgeting process.

Facility managers are increasingly including the environmental manager as a member of the capital project planning team. The environmental manager's involvement encourages early discussion of any environmental considerations relating to the capital project and ensures that permitting schedules and costs are included in the capital plan. A six-month delay in beginning construction of a multimillion dollar expansion project can literally add millions of dollars to the project cost. For instance, if major equipment has been ordered and delivered prior to the time environmental approvals have been granted, working capital can be tied up for the duration of the delay. The key to managing the environmental aspects of capital projects is for the environmental manager to get involved early, stay involved, and assertively communicate the realities of environmental permitting and compliance to project and operation managers.

25.8 ENVIRONMENTAL COMMUNICATIONS

The best-laid environmental management plans can go astray if the relevant information is not properly communicated to all of the interested parties. The successful environmental manager must develop and maintain good relationships with the environmental agencies that have jurisdiction over the facility. Communications with the agencies should be forthright, open, and frequent and designed to foster a high degree of credibility. Chapter 28 explores the subject in further detail and offers several simple recommendations for developing and maintaining good relations with regulatory agencies.

Today's environmental manager must also be able to communicate effectively with several other major audiences. First, the facility production managers must be initially educated and continuously informed of environmental requirements applicable to the facility. Most successful environmental managers participate in periodic production staff meetings and frequently visit production managers' offices to discuss environmental compliance and management issues. Understanding the needs and problems of production managers and providing them with ideas for maintaining environmental compliance while achieving production goals are among the most important responsibilities of the facility environmental manager.

Good communication with the nonmanagerial work force is also important. These employees often have primary responsibility for collecting environmental data or samples, operating pollution control equipment, or responding to spill events. They need to be fully informed, trained, and aware of their environmental responsibilities. They also need to feel free to contact the environmental manager regarding environmental issues or upsets. If your facility has a periodic newsletter or bulletin board, let everyone in the facility know about environmental improvements, pollution prevention possibilities, environmental incidents, and any other information that will allow the facility to maintain a high level of environmental awareness.

Finally, environmental managers should have strong external communication skills and the ability to discuss facility environmental issues with concerned citizens, environmental groups, and the news media. Contacts with these groups also create opportunities to develop rapport, which can be used to the advantage of the facility in times of need. A facility with a progressive environmental management program usually benefits by sharing the details of its program with the public.

Depending on the size of the facility, external communications may become a team effort. In a large facility, the plant manager, communications manager, and the environmental manager are involved in providing environmental information to the public. When more than one person represents the facility on such issues, it is essential to coordinate with facility managers to ensure a consistent company response.

25.9 CITIZEN INVOLVEMENT

The nation's environmental laws provide for citizen participation, and, as these laws are amended, the trend is to increase the level of citizen involvement. In the past, citizen involvement in environmental matters was traditionally limited to complaints by neighbors or comments at public hearings. Today it is common for citizen groups to propose legislation and attempt to influence the policies and practices of regulatory agencies. Citizens also have significant rights to review agency files for environmental data and facility reports and are able to file lawsuits if they can prove the authorized regulatory agencies have not properly addressed violations of environmental laws. Typically, these suits seek to recover enforcement penalties and legal fees and to force ''donations'' by the facility to a worthy environmental cause.

The citizen suit is a powerful tool in that it has undermined the principle of agency discretion in enforcement and requires facilities to be even more diligent about exceedances of permit conditions. Because citizen suits are usually based on evidence of noncompliance provided to the agency by the facility, these cases are difficult to defend and hard to explain to your neighbors and investors.

One way of helping to avoid citizen suits is to develop a formal program for responding to public concerns and citizen complaints. The program should specify procedures that describe how complaints or inquiries are to be handled and who is to provide the company response. When complaints or inquiries are received, this information should be recorded, and the facility should periodically review these records to determine whether any trends or patterns of complaints are developing. Citizen concerns can often be diffused at an early stage if complaints are promptly and thoughtfully addressed.

25.10 CULTURE AND INTEGRATION

Good records, timely reporting, and well-managed internal and external communications will go a long way toward keeping your facility in compliance with the environmental laws. However, the difference between a good environmental management program and a great one relates to the culture of the particular facility and how environmental management is integrated into the daily operation of the business. Are environmental upsets addressed as priorities, or are they resolved when such attention conveniently fits into the plant production schedule? Is environmental compliance one of the performance criteria of facility production managers? Is environmental compliance one of the primary objectives of the business, or is it considered an ''add-on'' responsibility?

A quick review of this book underscores the immensity of the task and leads to the conclusion that environmental compliance cannot be solely the responsibility of the environmental manager. Successful facilities recognize this fact and integrate environmental compliance responsibilities into all of the line manager positions. The solution to environmental violations is not to continue adding environmental staff to report incidents and upsets but, rather, to eliminate such incidents. The most profitable and well-run industrial plants are normally the facilities with the best environmental performance record, inasmuch as environmental upsets are usually good indicators of production problems or maintenance issues. A strong environmental compliance program will not only protect a facility from environmental liabilities but can add to the bottom line by reinforcing the importance of conducting necessary maintenance and avoiding serious production problems.

25.11 OTHER KEY AREAS OF ENVIRONMENTAL MANAGEMENT

In this chapter we have identified the basic areas of environmental management that should be included in a facility environmental management plan. Additional elements of the plan involve specific management requirements for the basic environmental media: water, air, solid/hazardous waste, and chemical components. Other sections of this book are devoted to the legal requirements of each specific regulatory program, and these chapters should also be consulted in evaluating the elements of a successful environmental management program.

26

ENVIRONMENTAL MANAGEMENT AND COMPLIANCE OVERSIGHT PROGRAMS

Compliance oversight programs are intended to ensure that operating facilities are actually meeting their environmental compliance and management obligations. This chapter summarizes three tools that corporate managers can use to meet this objective: the environmental compliance audit, the environmental management self-assessment program, and the comprehensive, compliance-focused environmental management system.

26.1 ENVIRONMENTAL COMPLIANCE AUDITS

Manufacturing facilities are responsible for complying with an ever-increasing number of complex environmental laws, and regulations and regulatory agencies have become much more aggressive in assessing significant civil and criminal penalties for violating the nation's environmental laws. Facility managers and company officers can be held personally liable for knowing and willful violations of the law. As a result of these changes in the regulatory world, environmental managers must ensure that their facilities comply with all applicable environmental legal requirements. Environmental compliance auditing is a basic tool used by many companies to identify and address environmental compliance issues.

An environmental compliance audit is a systematic and objective evaluation of a facility's compliance with environmental legal requirements. Before a company starts an environmental audit program, there should be a commitment by the company's top management to address promptly any compliance issues identified during the course of an audit. Without this commitment in advance, an environmental audit should not be conducted at the facility. Otherwise, unaddressed issues can evolve into allegations of known violations of environmental legal requirements and provide a basis for criminal liability.

An effective environmental audit program can enable companies to meet the following objectives:

- Identify environmental compliance issues and promptly remedy them.
- Increase employee awareness of environmental responsibilities.
- Anticipate regulatory inspections and thereby reduce regulatory enforcement actions and penalties.
- Enhance the company's credibility with its employees, customers, shareholders, and the community.

There are numerous opinions as to how best to conduct an environmental audit. The first step is to identify the environmental regulatory programs that will be considered within the scope of the audit. Typically, an environmental audit will examine regulatory compliance relating to some or all of the topics listed in Table 26.1.

A wide range of environmental audit checklists and questionnaires are used by environmental audit teams. Ideally, any checklist used by an audit team should be customized to reflect the scope of the audit and the practices of the particular industry. By way of example, a very general environmental audit checklist is provided in Appendix D.

Once the scope of the audit has been defined, the environmental audit

Table 26.1 Environmental compliance audit topics

- Air pollution
- Water pollution
- Solid waste management
- Hazardous waste management
- Hazardous material and petroleum storage and management
- Underground storage tanks
- TSCA[a] and PCB[b] management
- Safe Drinking Water Act
- Emergency Planning and Community Right-to-Know Law
- Emergency management programs

[a]Toxic Substances Control Act
[b]Polychlorinated biphenyl

team is assembled. It is important for the audit team to be independent of the audited facility and for audit team members to understand both the environmental laws and the practices of the affected industry. Although many states have enacted environmental audit privilege statutes that provide a qualified privilege for environmental audit reports, these statutes likely will not protect audit reports from disclosure in federal proceedings. Accordingly, in most cases it is advisable to include an in-house counsel or outside attorney on the audit team and involve the attorney in the preparation of the audit report for the sake of asserting the attorney-client privilege.

Prior to beginning any audit, the audit team should inform the facility manager of the schedule, nature, and scope of the audit and what is expected of the facility management. The audit team should obtain copies of all environmental permits, inspection reports, internal environmental reviews, and other related materials from the facility before conducting the audit site visit.

The site visit portion of the audit should begin with an initial meeting, introducing the audit team and describing the audit process. This kickoff meeting will be followed by tours of the facility, a review of the environmental permits and records, inspections of various facility operating practices, and detailed discussions of the facility's environmental compliance issues with key facility managers. Daily morning briefings with those managers, to summarize the findings of the previous day, also are recommended.

At the end of the site visit, the leader of the audit team should summarize for facility management the list of identified compliance issues and work with the facility to develop an "action plan" designed to promptly resolve these compliance matters. The action plan should include an implementation schedule and assign responsibilities for each task within the plan. Once a satisfactory plan has been developed, the audit team should schedule periodic reviews with the facility to ensure action plan items are addressed in a timely manner. When all action items have been properly addressed, the audit can be formally closed.

Audits should be conducted periodically, because both environmental laws and a facility's operations are constantly changing. Audit team members should also be rotated periodically to ensure an infusion of new ideas and perspectives into the audit process.

26.2 ENVIRONMENTAL MANAGEMENT ASSESSMENTS

In addition to conducting periodic environmental compliance audits, facilities should consider evaluating the strengths of their environmental management programs. The Global Environmental Management Initiative (GEMI) organization and other groups have established self-assessment guidelines for evaluating environmental management programs. Self-assessment programs allow facilities to continuously improve the quality of their environmental management programs while raising the profile of environmental compliance among all company employees.

The GEMI Environmental Self-Assessment Program is a structured approach that evaluates a facility's environmental management performance relative to the 16 principles for environmental management established by the International Chamber of Commerce (ICC). The ICC principles are briefly summarized as follows:

International Chamber of Commerce
Business Charter for Sustainable Development
Principles for Environmental Management

1. *Corporate Priority.* Environmental management should be recognized as one of the highest corporate priorities.
2. *Integrated Management.* Environmental management considerations should be integrated into all management functions of the organization.
3. *Process of Improvement.* Companies should continuously improve corporate policies, programs, and environmental performance and apply the same environmental criteria internationally.
4. *Employee Education.* Companies should educate and train employees to conduct their activities in an environmentally responsible manner.
5. *Prior Assessment.* Environmental impacts should be assessed before starting a new activity or project.
6. *Products and Services.* Companies should develop and provide products or services that have no undue environmental impact and are efficient in their consumption of energy and natural resources.
7. *Customer Advice.* Companies should advise and educate customers and the public in the safe use, transportation, storage, and disposal of products.
8. *Facilities and Operations.* Companies should develop, design, and operate facilities to minimize adverse environmental impacts and encourage the efficient use of energy and materials.
9. *Research.* Companies should research how to minimize the adverse environmental impacts of their operations.
10. *Precautionary Approach.* Companies should manufacture, market, and use products designed to prevent serious or irreversible environmental degradation.

11. *Contractors and Suppliers*. Companies should encourage their contractors and suppliers to adopt the same environmental management principles implemented by the companies themselves.
12. *Emergency Preparedness*. Companies should develop and maintain, where appropriate, emergency preparedness plans.
13. *Transfer of Technology*. Companies should participate in the transfer of environmentally sound technology and management methods.
14. *Contributing to the Common Effort*. Companies should contribute to the development of public policies and initiatives that will enhance environmental awareness and protection.
15. *Openness to Concerns*. Companies should encourage dialogue with employees and the public regarding their concerns about the environmental impacts of company facilities.
16. *Compliance and Reporting*. Companies should measure environmental performance and conduct regular environmental audits and assessments of compliance with company requirements.

More information regarding the detailed Environmental Self-Assessment Program is available from GEMI by contacting its office in Washington, D.C.

26.3 COMPLIANCE-FOCUSED ENVIRONMENTAL MANAGEMENT SYSTEMS

In August 1997, the Environmental Protection Agency's (EPA) National Enforcement Investigation Center (NEIC) published its guidance regarding the elements of a comprehensive, compliance-focused environmental management system (EMS).

This compliance-focused EMS model has been used as the basis for EMS requirements in several settlement agreements negotiated between EPA and offending companies. The guidance document should be of interest to all environmental managers, as it outlines EPA's expectations regarding the key elements of a successful EMS. In addition to following the GEMI Environmental Self-Assessment Program, environmental managers should consider evaluating their management programs relative to meeting the requirements of the NEIC guidance. The compliance-focused EMS provisions are summarized as follows:

1. *Management Policies and Procedures*
   a. *Environmental Policy*. The policy must clearly communicate management's commitment to environmental compliance.

b. *Site-Specific Environmental Policies and Standards*. The facility should identify policies, rules, and procedures, including standard operating procedures for each site. These procedures should clearly identify organizational responsibilities for maintaining regulatory compliance and the means of communicating environmental information to all organizational personnel and outside contractors. Procedures should also be developed to encourage interaction with regulatory agencies, and within the organization, regarding environmental requirements and regulatory compliance.

2. *Organization, Personnel, and Oversight of EMS*. The plan should explain how the EMS is implemented and maintained. A portion of the plan should include organizational charts that identify and define duties, roles, and responsibilities of key environmental program personnel in implementing the EMS.
3. *Accountability and Responsibility*. The EMS should specify who is responsible for environmental compliance within specific portions of the facility and explain the consequences for violating specified operating procedures.
4. *Environmental Requirements*. The EMS should identify, interpret, and communicate environmental requirements to employees, on-site service providers, and contractors.
5. *Assessment, Prevention, and Control*. The EMS must identify an ongoing process for preventing and controlling releases, ensuring environmental protection, and maintaining compliance with environmental requirements. The EMS should describe the monitoring and measurements that are required to ensure sustained compliance. The facility must also describe the procedures for identifying the activities that could result in adverse environmental impacts and develop a system for routine, objective, environmental self-inspections. Finally, the facility should have a process for incorporating environmental requirements into the planning, design, and operation of the facility.
6. *Environmental Incident and Noncompliance Investigations*. The EMS must establish procedures for investigating environmental incidents and noncompliance situations, and verify that appropriate corrective actions have been taken.
7. *Environmental Training, Awareness, and Competence*. The EMS must include programs for training employees to ensure that they are aware of any applicable environmental requirements and procedures. Environmental managers, in particular, must have enough education, training, and/or experience to ensure a basic level of competence.

8. *Planning for Environmental Matters.* The EMS must explain how environmental planning is integrated into the facility's business planning process.
9. *Maintenance of Records and Documentation.* The EMS should identify the records that are required to support the system and specify the facility's environmental data management systems.
10. *Pollution Prevention Program.* The EMS must include an internal program for preventing, reducing, recycling, reusing, and minimizing waste and emissions.
11. *Continuing Program Evaluation and Improvement.* The EMS must be evaluated periodically (at least annually) to incorporate any required program improvements.
12. *Public Involvement/Community Outreach.* The facility must have a program for ongoing community education and involvement regarding environmental impacts of the facility's operations.

The GEMI Environmental Self-Assessment Program and the NEIC compliance-focused EMS requirements both recognize one of the absolute truths of environmental compliance management: that is, failure to comply with applicable environmental requirements is often the result of a poorly designed and executed environmental management program. Implementation of a reliable environmental management system and organization will not only improve compliance and reduce the facility's exposure to environmental penalties, but may also allow the organization to reduce the costs of maintaining compliance.

27

INTEGRATING ENVIRONMENTAL AND SAFETY MANAGEMENT

Statute:	(''OSHA'') 29 USC § 651 et seq.
Regulations:	29 CFR Parts 1902 to 1990

Although pollution control and worker safety issues are governed by completely different statutory and regulatory programs, the legal requirements of these programs have begun to overlap, especially in regard to chemical management and regulating the releases of hazardous substances. Indeed, in some instances the proper administration of a worker safety program may be critical to ensure a facility's environmental compliance. A working knowledge of the Occupational Safety and Health Act (OSHA) is essential for any environmental manager. The purpose of this chapter is to provide an overview of OSHA and a summary of two of the worker safety programs that relate to environmental compliance.

27.1 OVERVIEW OF THE OCCUPATIONAL SAFETY AND HEALTH ACT

27.1.1 Background

The Occupational Safety and Health Act of 1970 is designed to provide safe and healthful working conditions for all American workers. The Act requires employers to furnish their employees with a safe work environment and authorizes stiff civil and criminal penalties for noncompliance. The Act is administered federally by the Department of Labor through the Occupational Safety and Health Administration. States can be delegated authority to administer the OSHA requirements, and approximately half of the states administer their own programs. Once a state plan is approved by the federal OSHA, there typically is little federal involvement in the state's administration of the program.

27.1.2 To Whom Does the Act Apply?

OSHA regulates virtually every person and business organization that hires employees. Congress intended the coverage of the Act to be as broad as the scope of the commerce clause of the Constitution, declaring it to be applicable to every person "engaged in a business affecting commerce who has employees." The Act does have some exclusions and modifications, however, for businesses regulated under the Atomic Energy Act, under job safety rules of federal agencies other than the Department of Labor, and for government employees.

27.1.3 General Provisions of the Act

OSHA Standards. The OSHA standards are the "heart" of the Act—the regulations that employers must observe. Penalties and abatement orders can be imposed for violations of these standards.

The OSHA standards are divided into three general categories: construction standards, maritime standards, and general industry standards. There is no consistency in the way OSHA categorizes its standards, and employees engaged in the construction and maritime industries also are governed by many general industry standards.

The General Duty Clause. The general duty clause of OSHA provides that each employer "shall furnish to each of his employees employment and a place of employment which are free from recognized hazards that are causing or are likely to cause death or serious physical harm to his employees." The intent of the general duty clause is to fill any voids that may exist in the OSHA standards. OSHA inspectors tend to use the general duty clause to establish ad hoc standards when they cannot identify an OSHA standard that regulates the condition or practice that they think should be addressed.

OSHA Record Keeping. The OSHA record-keeping regulations require nonexempt employers to maintain a log and summary of all "reportable occupational injuries and illnesses." OSHA recommends that employers use

OSHA Form No. 200 for recording these injuries and illnesses. OSHA has promulgated regulations that provide assistance to employers in the record-keeping process, but these guidelines can be confusing, vague, and ambiguous. During its regulatory inspections, OSHA reviews these injury/illness records and issues penalties for companies that fail to comply with the OSHA record-keeping requirements.

27.2 OSHA HAZARD COMMUNICATION STANDARD

The OSHA hazard communication standard is designed to inform workers of the existence of potentially dangerous substances in the workplace and how to protect themselves against those dangers. The hazard communication standard generally provides that chemical manufacturers and importers must assess the hazards of the chemicals they produce or import and furnish detailed information to their customers regarding any hazardous chemicals. Thereafter, all employers must provide that information to their employees by means of a hazard communication program, labels on containers, material safety data sheets, and access to written records and documents.

The major elements of the hazardous communication standard include:

- Chemical hazard determinations,
- A written hazard communication program,
- Employee information and training,
- Material safety data sheets (MSDSs), and
- Container labeling.

Compliance with the requirements of the OSHA hazard communication program and the proper management of MSDSs are required in order for facilities to meet their legal obligations under the Community-Right-to-Know Law. The regulatory requirements for each of the five major elements of the program are summarized in the following paragraphs.

27.2.1 Hazard Determination

Chemical manufacturers and importers are required to evaluate the chemicals and mixtures of chemicals they produce or import so as to determine whether they are hazardous chemicals. In general, a hazardous chemical is any chemical that presents either a ''physical hazard'' or a ''health hazard.'' A chemical's physical and health hazard data must be evaluated in accordance with detailed procedures specified in the OSHA regulations.

27.2.2 Written Hazard Communication Program

Employers must develop, implement, and maintain at the workplace a written hazard communication program. This program must include the following:

- A list of hazardous chemicals in the workplace or in each work area,
- The methods the employer will use to inform employees of the hazards associated with these chemicals, and
- A description of how the labeling, MSDS, and employee training requirements will be met.

Employers who produce, use, or store hazardous chemicals at a workplace in such a way that employees of outside contractors may be exposed must also include the following information in the written program:

- How MSDSs will be made available to the outside contractor for each hazardous chemical,
- How the employer will inform the outside contractor of precautions necessary to protect the contractor's employees both during normal operating conditions and in foreseeable emergencies, and
- The methods the employer will use to inform contractors of the labeling system used in the workplace.

The written hazard communication program must be made available upon request to employees, their designated representatives, OSHA, and the National Institute for Occupational Safety and Health (NIOSH).

27.2.3 Employee Information and Training

Employers must provide their employees with information and training in regard to hazardous chemicals in their work areas at the time of their initial assignment and whenever a new hazard is introduced into a work area. Training must be sufficient to cover the volume and complexity of the hazardous chemicals in the workplace. Although no specific requirement for recurring training is given, the regulations stipulate that supplemental training must be provided whenever a new hazard is introduced into a work area.

In general, employees must be informed of

- The requirements of the hazard communication program,
- The location of hazardous chemicals in the work area, and

- The location and availability of the written hazard communication program, including the required list of hazardous chemicals and MSDSs.

Employee training must include

- Methods and observations to detect the presence or release of a hazardous chemical in the work area,
- The physical and health hazards of the chemicals in the work area,
- Personal protective measures, including emergency procedures, and
- The details of the hazard communication program developed by the employer, including the labeling and MSDS systems and how to obtain and use appropriate hazard information.

27.2.4 Material Safety Data Sheets (MSDSs)

Employers must have on file an MSDS for each hazardous chemical used in the workplace. Although no specific form is required by the regulations, MSDSs must be in English and must contain, at a minimum, the following information:

- If the hazardous chemical is a single substance, its chemical and common names
- If the hazardous chemical is a mixture that has been tested as a whole, the chemical and common names of the ingredients that contribute to the hazards and the common name of the mixture itself
- If the hazardous chemical is a mixture that has not been tested as a whole, the chemical and common names of the ingredients that have been determined to be health hazards
- The physical and chemical characteristics of the hazardous chemical (such as flash point, vapor pressure)
- The physical hazards of the chemical, including the potential for fire, explosion, and reactivity
- The health hazards of the chemical, including signs and symptoms of exposure and any medical conditions that are generally recognized as being aggravated by exposure to the chemical
- The primary routes of entry to the body
- The OSHA Permissible Exposure Limit and any other exposure limit used or recommended by the manufacturer, importer, or employer

- Whether the chemical is listed in the National Toxicology Program's latest report on carcinogens or has been found to be a potential carcinogen
- Any generally applicable precautions for safe handling and use that are known to the manufacturer, importer, or employer, including hygiene practices, protective measures, and spill cleanup procedures
- Any generally applicable control measures that are known to the manufacturer, importer, or employer, such as appropriate engineering controls, work practices, or personal protective equipment
- Emergency and first aid procedures
- The date the MSDS was prepared or the date of the latest change to it
- The name, address, and telephone number of the chemical manufacturer, importer, employer, or other responsible party who prepared the MSDS and can provide additional information on the chemical and appropriate emergency procedures.

The MSDS preparer must ensure that the information on the MSDS is supported by the scientific evidence used in making the hazard determination. When the MSDS preparer becomes aware of significant new information regarding the MSDS, it must be added to the MSDS within three months.

Employers must maintain current copies of the MSDS for each hazardous chemical in the workplace and must ensure that they are readily accessible to employees in the work area during each work shift. Where an employee's duties include work at different geographic locations, MSDSs may be centrally located, but employees must be able to obtain emergency information immediately. MSDSs must also be available for review by OSHA and designated employee representatives.

27.2.5 Container Labeling

Chemical manufacturers, importers, and distributors must ensure that all containers of hazardous chemicals leaving the workplace are labeled or marked, showing

- The identity of the hazardous chemicals,
- Appropriate hazard warnings, and
- The name and address of the manufacturer, importer, or other responsible party.

The employer is responsible for ensuring that each container of hazardous chemicals in the workplace is labeled with

- The identity of the hazardous chemical in the container and
- The appropriate hazard warnings.

Labels or other forms of warning on a container must be legible, in English, and prominently displayed on the container or readily available in the work area throughout each work shift. When appropriate, the required information may be displayed in a foreign language in addition to English. Employers must also ensure that labels are not defaced or removed from containers in the workplace.

Certain substances are exempt from the labeling requirements of the hazard communication program.

27.3 PROCESS SAFETY MANAGEMENT OF HIGHLY HAZARDOUS CHEMICALS

OSHA's process safety management (PSM) regulations for highly hazardous chemicals became effective in May 1992. The federal regulations are designed to prevent or minimize catastrophic releases of chemicals that could result in toxic, fire, or explosion hazards. All processes involving listed chemicals that exceed specified threshold quantities and certain large quantities of flammable liquids or gases are governed by the regulation. For example, the following commonly used chemicals and threshold quantities have been listed by OSHA:

- Chlorine (1,500 lbs)
- Liquid sulfur dioxide (1,000 lbs)
- Chlorine dioxide (1,000 lbs)
- Hydrogen sulfide (1,500 lbs)
- Anhydrous ammonia (10,000 lbs)

Many of the obligations imposed by the PSM regulations also meet the requirements established by the Enviromental Protection Agency (EPA) in its Risk Management Program (RMP) rules.

The major requirements of the PSM regulations are summarized in the following paragraphs.

27.3.1 Employee Participation

Employers must prepare a written plan providing both a consultation with employees on the development of process hazard analyses and other provisions of the standard, and access by employees to all information required by the standard.

27.3.2 Process Safety Information

Employers must prepare written process safety information pertaining to the hazards of chemicals used or produced in a process, the hazards in the technology of the process, and the hazards of the equipment used in the process.

27.3.3 Process Hazard Analysis (PHA)

Employers must prepare Process Hazard Analyses (PHAs) that identify, evaluate, and control hazards involved in each process. PHAs must be updated every five years, and employers must use a listed methodology or an appropriate alternative.

27.3.4 Operating Procedures

Employers must prepare written operating procedures for each regulated process, addressing the following elements:

- Steps for each operating phase,
- Operating limits,
- Safety and health considerations, and
- Safety systems and their functions.

Annual certification is required. Employers must also develop and implement safe work practices governing procedures such as lockout/tagout, confined space entry, opening process equipment, and piping and control over entrance into the facility by maintenance, contractors, and other support personnel.

27.3.5 Training

The training requirements of the federal PSM regulations are as follows:

- *Initial training*. Employees must be trained in process overview and operating procedures before being involved in a newly assigned process.
- *Refresher testing*. Requires employee training at least every three years.
- *Training documentation*. Requires employers to document that employees have received any required training.

27.3.6 Contractors

Employers must when selecting a contractor, consider the contractor's safety record, inform the contractor of hazards related to the process, periodically evaluate the contractor's compliance with these obligations, and maintain an illness/injury log for the contractor's employees. Contractors must train their employees in safe work practices, document that such training has been received and understood, be responsible for ensuring that their employees follow the facility's safety rules, and advise the facility employer of any hazards discovered by the contractor's employees.

27.3.7 Prestart-up Safety Review

Employers must perform a prestart-up safety review for new facilities and modified facilities to which there is significant modification. This review should include adequacy of construction and equipment; safety, operating, maintenance, and emergency procedures; employee training; and completion of PHAs.

27.3.8 Mechanical Integrity

Employers must establish written procedures for maintaining the ongoing integrity of process equipment, training of maintenance employees, inspection and testing consistent with good engineering practices, and documentation of inspections and tests.

27.3.9 Hot Work Permit

Employers must issue permits for hot work operations (e.g. welding, brazing, etc.) on or near a regulated process.

27.3.10 Management of Change

Employers must establish written procedures for managing changes to process chemicals, technology, equipment, procedures, or facilities. Maintenance and contractors' employees affected by the change must be trained. Process safety information and operating procedures must be updated if applicable.

27.3.11 Incident Investigation

Employers must investigate within 48 hours of "each incident which resulted in, or could reasonably have resulted in, a catastrophic release of highly hazardous chemicals in the workplace.'' Employers must promptly implement any recommendations and document the facility's response.

27.3.12 Emergency Planning and Response

Employers must implement an emergency action plan for the entire plant. The plan must include procedures for handling small releases.

27.3.13 Compliance Audits

Employers must evaluate compliance with the PSM standard at least every three years. An audit report must be prepared documenting the evaluation and describing the action taken to address the identified deficiencies.

27.3.14 Trade Secrets

Employers must make available to employees and employee representatives all information necessary to comply with the PSM standard, without regard to trade secret status. The employer may require these individuals to enter into confidentiality agreements.

27.4 MANAGEMENT PROCEDURES

The facility's environmental manager and safety manager should work together as a team to ensure all environmental and safety requirements are met. Many of the planning, training, and implementation efforts should be coordinated between the two managers to ensure a consistent approach within the facility and avoid duplication of efforts. The managers should also keep each other advised of new regulatory programs and discuss what, if any, impact the new rules will have on both the environmental and safety management programs. Most facilities are not organized in such a way that this coordination will happen naturally. Thus, the burden typically falls on the environmental and safety managers to overcome institutional obstacles and provide the facility with a well-coordinated environmental and safety management program.

28

MANAGING RELATIONSHIPS WITH AGENCIES

The environmental laws and regulations in the United States are implemented and administered by government agencies at the federal, state, and local levels. These environmental agencies are the environmental manager's major point of contact in regard to managing a facility's environmental compliance program. It is important to develop a detailed understanding of these environmental agencies and maintain excellent rapport with the agency representatives involved with the facility.

Environmental agencies have the lead role in developing new programs to regulate a facility's operations and deciding whether to initiate enforcement actions. The development of new regulations is a public process in which both concerned citizens and the affected industry can participate. Initial drafts of proposed regulations often contain language or conditions that may adversely affect the way your facility operates or impose economic hardship on the facility. The environmental manager who has established credibility with the environmental agency stands a much better chance of affecting the outcome of a rule-making proceeding. Credibility is also a key factor in negotiating permit terms with agency representatives or convincing an agency that an enforcement action is not warranted in a particular case. This type of credibility cannot be quickly or easily established and is usually the result of years of forthright dialogue between agency and facility representatives.

28.1 ENVIRONMENTAL AGENCY OVERVIEW

The Environmental Protection Agency (EPA) is the federal agency responsible for administering the federal environmental laws and regulations. Each state, in turn, has established an agency designed to implement the federal programs and any additional state environmental laws or regulations. However, not all states are delegated by EPA the full authority to administer the federal environmental programs, which often leads to much confusion regarding the respective roles of EPA and the states.

Some states have decided to delegate their responsibilities to completely independent local agencies. For instance, the state of California has established the South Coast Air Quality Management District to administer Clean Air Act programs in the Los Angeles area, with the state environmental agency playing only an oversight role. This layering of responsibilities creates the very real possibility that a single facility could be subject to several different regulatory programs administered by different agencies. Obviously, an environmental manager must learn what agencies have what responsibilities within this potentially confusing maze of overlapping regulatory authority.

28.1.1 The Environmental Protection Agency

The Environmental Protection Agency, established in 1970 by Congress, is headquartered in Washington, D.C., but has established offices, laboratories, and research centers throughout the country. EPA maintains a website located at http://www.epa.gov/, which lists the various agency offices and their purposes. EPA has developed the following mission statement:

> **EPA's Mission**
>
> The mission of the **U.S. Environmental Protection Agency** is to protect human health and to safeguard the natural environment—air, water, and land—upon which life depends.
>
> EPA's purpose is to ensure that:
> - All Americans are protected from significant risks to human health in the environment where they live, learn, and work.
> - National efforts to reduce environmental risk are based on the best available scientific information.
> - Federal laws protecting human health and the environment are enforced fairly and effectively.
> - Environmental protection is an integral consideration in U.S. policies concerning national and international trade, and these factors are similarly considered in establishing environmental policy.
> - All parts of society—communities, individuals, business, state and local governments, tribal governments—have access to accurate information sufficient to effectively participate in managing human health and environmental risks.

- Environmental protection contributes to making our communities and ecosystems diverse, sustainable, and economically productive.
- The United States plays a leadership role in working with other nations to protect the global environment.

Has EPA achieved its mission? There is little doubt that there have been significant improvements in this country's environmental quality since 1970, especially in the areas of conventional air emissions, water discharges, and solid and hazardous waste management. On the other hand, increased public expectations, population growth, and increased scientific knowledge are creating new challenges for the agency. For the industrial environmental manager, EPA will be an important part of life for the foreseeable future, although it is likely that EPA's traditional ''command-and-control'' approach to environmental regulation will evolve into a more cooperative partnership with American industry.

EPA's national programs are aimed at developing policy and new regulations, expanding the scope of scientific knowledge, and coordinating nationwide enforcement efforts. Most industrial environmental managers will spend little time working with EPA on its national programs, other than possible involvement in agency rule making or questions regarding the agency's analytical testing methods. Rather, contact with EPA will most often occur at the EPA regional office level.

EPA regional offices implement and execute federal programs involving the states (and territories) in their regions. There are 10 EPA regional offices displayed on the map in Figure 28.1. The states associated with each regional office also are listed in the figure. An environmental manager's involvement with an EPA regional office typically relates to permitting of new sources or renewal of permits in states without delegated authority, enforcement issues, facility inspections, and general requests for regulatory guidance. Even in states with delegated authority for permit programs, the EPA regional office has oversight and review authority and often becomes significantly involved in the review and approval of environmental permits.

28.1.2 State and Local Environmental Agencies

State and local environmental agencies were created to implement and develop state environmental policies and regulations. Although state environmental regulations usually mirror the federal requirements, states are authorized to enact programs that go far beyond the federal requirements. If a state's or local agency's regulatory program is deemed to meet federal requirements, then EPA can delegate the administration of the program to the state or local agency. Even without federal delegation, a state or local agency

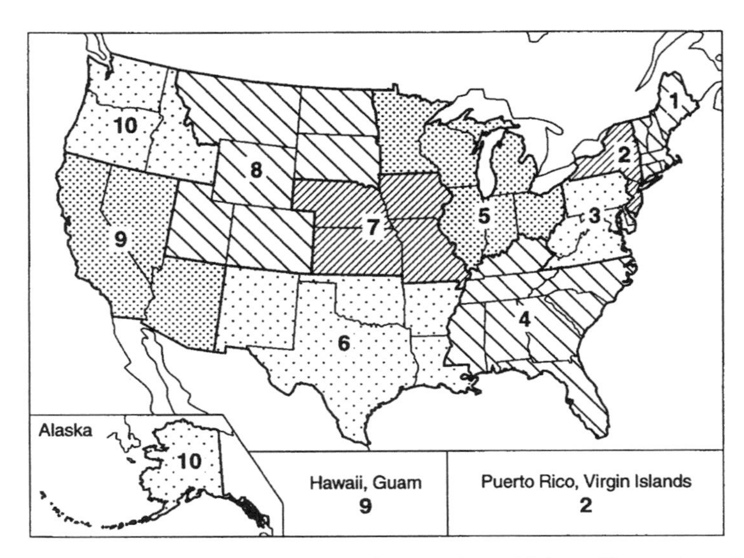

Figure 28.1 EPA Regions. *Region 1*: Connecticut, Maine, Massachusetts, New Hampshire, Rhode Island, and Vermont. EPA regional office located in Boston, Massachusetts (617-565-3400). *Region 2*: New Jersey, New York, and territories of Puerto Rico and the U.S. Virgin Islands. EPA regional office located in New York, New York (212-637-3000). *Region 3*: Delaware, Maryland, Pennsylvania, Virginia, West Virginia, and the District of Columbia. EPA regional office located in Philadelphia, Pennsylvania (215-566-5000). *Region 4*: Alabama, Florida, Georgia, Kentucky, Mississippi, North Carolina, South Carolina, and Tennessee. EPA regional office located in Atlanta, Georgia (404-562-9900). *Region 5*: Illinois, Indiana, Michigan, Minnesota, Ohio, and Wisconsin. EPA regional office in Chicago, Illinois (312-353-2000). *Region 6*: Arkansas, Louisiana, New Mexico, Oklahoma, and Texas. EPA regional office located in Dallas, Texas (214-665-2200). *Region 7*: Iowa, Kansas, Missouri, and Nebraska. EPA regional office located in Kansas City, Kansas (913-551-7003). *Region 8*: Colorado, Montana, North Dakota, South Dakota, Utah, and Wyoming. EPA regional office located in Denver, Colorado (303-312-6312). *Region 9*: Arizona, California, Hawaii, Nevada, and the territories of Guam and American Samoa. EPA regional office located in San Francisco, California (415-744-1500). *Region 10*: Alaska, Idaho, Oregon, and Washington. EPA regional office located in Seattle, Washington (206-553-1200).

may require a state permit or license for air emissions or water discharges in addition to any required federal permits.

State and local environmental agencies are almost always the main points of contact for facility environmental managers. These agencies are primarily responsible for ensuring that a facility complies with applicable federal environmental requirements. State environmental agencies also conduct periodic inspections of facilities, develop their own regulatory programs, and offer facilities technical advice and guidance.

28.2 DYNAMICS OF RELATIONSHIPS WITH AGENCIES

At most facilities, the environmental manager is the primary point of contact with the environmental agencies. Communications with the state and local agencies responsible for day-to-day administration of the facility's permits should be frequent, open, and honest. The environmental manager should develop and maintain a professional relationship with agency staff, because this relationship is one of the most important elements of a successful environmental management program. There undoubtedly will be times when the environmental manager will be discouraged by the weight and pace of the agency's bureaucracy. Yet the manager must always remember that agencies are bound by detailed regulatory requirements and must take into consideration the concerns of a broad spectrum of interested parties before rendering a decision. Environmental managers should also recognize that agencies are often underfunded and are subject to the same economic constraints faced by industry. Agencies are granted a great deal of latitude in interpreting the environmental laws and will often give a facility the benefit of the doubt in a questionable situation if it views the facility and its environmental manager as credible. Such credibility is not easily earned and can be destroyed by a single act of deception or continuing compliance problems. Environmental managers should keep their agencies informed of upset events at their facilities that could lead to regulatory violations or complaints from neighbors. An agency would much rather learn of these situations from the facility than from a disgruntled neighbor or member of the public. When in doubt, err on the side of notifying the agency. Cooperation with these agencies usually yields impressive dividends over the life of the facility.

28.3 TRADE ASSOCIATIONS AND GOVERNMENT AFFAIRS

When new environmental laws or regulations are proposed, the process is open to public comment through either the legislative or the rule-making process. Environmental managers have the right to either submit written comments or testify at legislative or regulatory hearings. This type of individual effort often may not be practical and, in regard to major regulatory proposals, may not be very effective.

Often the solution for American industry is to form a trade association and let the association speak for a large group of similarly situated facilities. Trade associations are active at the federal, state, and local levels of government. Examples of national trade associations are the American Petroleum Institute (API) and the Chemical Manufacturer's Association (CMA). Trade associations that are organized to influence environmental regulation typically assemble a group of environmental managers from different facilities to develop an opinion on a proposed new rule. Then the trade association provides consensus comments or testimony to the environmental agency, state legislature, or Congress on behalf of the affected facilities. These comments often result in modifications to the proposed statute or regulation or, at the very least, some consideration as to whether it can practically be applied to those facilities.

Trade associations can be powerful voices when they represent a large group of industry interests. They also offer a good mechanism for informing environmental managers of current legal and regulatory developments. But these groups are not without fault. Attending association meetings can be tedious if your facility is not affected by the specific proposal under consideration. Trade association groups may have difficulty reaching consensus on a controversial issue, and whatever consensus they reach can sometimes be characterized as the "lowest common denominator."

Overall, trade associations seem to be flourishing. Despite their flaws, they offer a convenient and cost-effective way to evaluate and influence the endless wave of environmental regulation. As an environmental manager, it is important for you to recognize the issues that are important to your facility and to avoid becoming unduly involved in association meetings and conference calls.

29

ENFORCEMENT

The command-and-control system of environmental regulation used by the federal government in the United States results in a strong adversarial relationship between the Environmental Protection Agency (EPA) and the regulated community. Private citizens are also empowered to enforce the environmental laws when federal and state authorities fail to do so. Although EPA is beginning to explore alternative methods of environmental regulation, the agency's reliance on strong enforcement is well established. This chapter explores the inspection authority granted to EPA by the federal environmental laws, discusses how to handle regulatory inspections, summarizes the civil and criminal sanctions used by EPA to penalize companies and individuals who violate the nation's environmental laws, and examines the fundamental principles governing environmental citizen suits. Environmental managers need to understand these procedures to protect both the company and its managers from the adverse consequences of environmental enforcement actions.

29.1 FEDERAL INSPECTION AUTHORITY

All of the major federal environmental statutes contain provisions granting EPA broad authority to conduct regulatory inspections. These statutory inspection and entry provisions are briefly summarized in the following paragraphs.

29.1.1 Clean Air Act

Section 114 of the Clean Air Act (CAA) outlines EPA's authority to enter facilities and conduct inspections to determine compliance with the requirements of the Act.

This section provides that EPA or its authorized representative, upon presentation of proper credentials, is authorized to enter a facility subject to the Act to conduct an inspection. These inspectors can sample any emissions the facility is required to sample by EPA, obtain copies of records the facility is required to keep, and inspect any required monitoring equipment or method.

29.1.2 Clean Water Act

Section 308 of the Clean Water Act (CWA) allows EPA or its authorized representative (including an authorized contractor acting as a representative of EPA), upon presentation of proper credentials, to conduct inspections of facilities subject to the Act.

The Act allows inspectors to copy any records the owner is required to establish and maintain relating to any applicable effluent limitations or to toxic, pretreatment, or new source performance standards. An inspector is also authorized to inspect any monitoring equipment or methods required to be installed, used, and maintained by the owner or operator of any point source for the purposes of sampling the effluent and monitoring the operation of the facility.

29.1.3 Resource Conservation and Recovery Act

Section 3007 of the Resource Conservational Recovery Act (RCRA) authorizes any officer, employee, or representative of EPA to conduct inspections of any person who generates, stores, treats, transports, disposes of, or otherwise handles hazardous waste. RCRA inspectors are allowed to inspect any facilities, equipment, practices, or operations regulated or required by a RCRA permit. An inspector can monitor or obtain samples of any hazardous waste or any waste containers or labeling. If the inspector takes any samples, he or she must give a receipt describing the samples. If the samples are analyzed, EPA must provide the facility with a copy of the results of the analysis. Section 3007(c) of RCRA provides for annual inspections of storage or disposal facilities that are owned or operated by a department, agency, or instrumentality of the federal government. Section 3007(d) provides for annual inspections of storage or disposal facilities operated by a state or local government. Section 3007(e) directs EPA to develop a program to thoroughly inspect every treatment, storage, or disposal facility no less often than every two years.

29.1.4 Emergency Planning and Community Right-to-Know Act

There are no provisions in the Emergency Planning and Community Right-to-Know Act (EPCRA) specifically authorizing EPA to conduct regulatory inspections regarding the Act. EPA can, in connection with its procedures for issuing administrative penalties, issue subpoenas for the attendance and testimony of witnesses and the production of relevant papers, books, or documents.

29.1.5 Toxic Substances Control Act

Section 11 of the Toxic Substances Control Act (TSCA) authorizes EPA and any duly designated representative to inspect any facility that is subject to the requirements of TSCA. Such an inspection may be made only upon presentation of appropriate credentials and a written notice to the owner or operator of a facility. Under TSCA, the inspection can address all things within the facility bearing on compliance with the TSCA requirements applicable to chemical substances and mixtures within the facility.

29.2 MULTIMEDIA INSPECTIONS

Multimedia compliance inspections are designed to determine a facility's compliance status in regard to all applicable environmental laws, regulations, and permits. EPA has decided to increase the number of multimedia inspections, as the agency believes these inspections optimize its enforcement resources.

Representatives of EPA's National Enforcement Investigations Center (NEIC) in Denver, Colorado, often are involved in EPA multimedia inspections. Representatives from the appropriate EPA region, state environmental agency, and local environmental authority may also serve as members of the inspection team. The primary objective of the NEIC is to provide expertise and resources for development and support of civil and criminal environmental enforcement cases. NEIC has established detailed procedures and protocols for conducting multimedia inspections.

Multimedia inspections typically involve a team of inspectors who spend several days at a facility. The inspectors tour the facility, question facility personnel, review environmental records, and conduct environmental sampling. The evidence collected during the inspection is typically used to support appropriate enforcement actions against the targeted facility.

29.3 HOW TO HANDLE REGULATORY INSPECTIONS

The federal EPA and companion state agencies have dramatically increased their enforcement of the nation's environmental laws. Companies are often at a loss as to how to handle an environmental inspection and, consequently, fail to follow many of the procedures that would help a company defend itself in an environmental enforcement proceeding. The purpose of this summary, therefore, is to offer some simple suggestions that will help environmental managers properly respond to environmental agency inspections.

The following principles should be carefully considered by any environmental manager facing a multimedia or other environmental agency inspection:

1. Designate a team of individuals within your facility who will be responsible for handling environmental inspections. Obviously, the larger your facility, the more people you should include on this team. Team members should be familiar with the basic elements of the applicable environmental laws and the procedures followed by environmental inspectors. Legal counsel should be included on the team and should be directly involved in any inspection regarding particularly sensitive issues or any criminal investigation.
2. Anticipate EPA inspections by properly organizing pertinent records. Records that are not normally subject to inspection (e.g., environmental audit reports, internal memoranda, privileged communications) should be physically separated from other environmental compliance documents.
3. When agency inspectors arrive at your facility, notify an appropriate company representative. The company should designate a person who will act as the company's primary contact with the inspectors. This person is usually the facility's environmental manager.
4. Once the designated environmental manager is present, insist on an opening conference with the EPA inspection team and discuss the following items:
    - Determine what kind of inspection is to be done and define the scope of the inspection. If EPA is conducting a criminal investigation, legal counsel should be contacted immediately.
    - Determine how many environmental inspection teams will be in the facility. Ensure that each EPA inspector or EPA inspection team is accompanied by a company representative.

- Explain to EPA the company's document production system—that any requests for documents are to be in writing and given to a designated individual who will log the request, obtain the documents, screen them for applicable privileges (e.g., attorney-client privilege), and provide the information to EPA. This procedure should be followed for all documents provided to EPA, and the facility should keep a full set of copies of all documents given to the agency.
- Inform EPA that if the inspection team will be taking photographs, a company representative will take identical photographs at the same time.
- Determine whether any compliance sampling will be done. If so, tell EPA that the company intends to duplicate all measurements and samplings.
- Discuss with the EPA inspectors procedures for identifying and protecting trade secrets or confidential business information.

5. During the actual inspection, cooperate with the EPA inspectors and answer specific questions, but do not offer any information that goes beyond the questions asked by the inspector. Take detailed notes regarding what the inspectors saw and said and the employees who were interviewed. Accompany the EPA inspectors at all times. Do not allow an inspector to go into the facility unescorted, and try to limit the inspectors to the scope of the inspection. If an inspector goes beyond the scope of the inspection, consider the following options:
   - Attempt to reach agreement with the inspector regarding the scope of the inspection.
   - Involve legal counsel.
   - Take detailed notes regarding the activities that go beyond the scope of the inspection.
6. If any violations are observed by the agency inspector, you should try to correct those violations in other areas of the facility before the inspector visits them.
7. Try to correct any violations identified by EPA before the end of the inspection. Tell the inspectors once the violations have been corrected.
8. Request a closing conference with the EPA inspectors before they leave the facility. A closing conference can be used to correct errors relating to the inspection and to learn whether EPA intends to issue any formal citation. If EPA announces its intention to initiate an enforcement action, you should contact legal counsel and discuss how best to proceed.

29.4 HOW TO HANDLE AN ENVIRONMENTAL CRIMINAL INVESTIGATION

Criminal investigations of environmental violations are becoming more common. If an inspection officer arrives at your facility and announces an intent to investigate criminal violations, call your legal counsel immediately and ask the inspector to wait until your counsel arrives at the scene. If the inspector does not have a search warrant, deny access and tell the inspector that he or she must obtain a search warrant before you will allow the investigation. If the inspector already has a criminal warrant and will not wait for your legal counsel to arrive, you should allow the inspector to conduct the search.

Once a criminal investigation begins, you should follow the inspector during the search and record all places searched and all items seized. By law you are not required to give consent to search areas beyond those allowed by the warrant, and it is prudent to review the warrant carefully to understand the scope of the search. Finally, and most important, be mindful of your constitutional right to remain silent. Criminal prosecutions may be brought against individuals as well as companies, and any information you provide to the criminal investigator can be used against either you or the company in subsequent legal proceedings.

29.5 FEDERAL CIVIL PENALTIES

All of the major federal environmental laws contain a variety of civil enforcement tools available to the regulatory agencies. Rather than include a detailed summary of these requirements in this book, we have generally summarized the civil penalty provisions of the five major federal environmental statutes in Table 29.1.

State environmental statutes also establish civil penalties and other forms of enforcement relief that are generally consistent with the federal requirements.

Environmental agencies bring civil enforcement actions to enforce compliance with applicable legal requirements, protect the public health and the environment, and deter future violations of the environmental laws. Formal enforcement actions are often resolved when the violator agrees to enter into a consent decree or stipulation agreement whereby the facility agrees to pay a negotiated penalty and implement whatever remedial measures are necessary to address the identified issues of noncompliance.

States usually calculate civil penalties with the use of a penalty matrix that incorporates a variety of factors into the penalty calculations. One of the major factors considered by the agencies is the "economic benefit" violators obtain from their noncompliance with the environmental laws. EPA has developed economic models, such as the "BEN model," that are used to calculate the economic benefit of noncompliance. The BEN model estimates economic benefit by comparing two cases. The "on-time case" evaluates the cash flows that would have been necessary to achieve timely compliance. The "delay case" considers the cash flows actually incurred to achieve delayed compliance. The difference between the two cases is the "economic benefit" of noncompliance.

Another concept frequently used in civil penalty negotiations is a request to mitigate civil penalties in exchange for environmental enhancement projects, which are commonly referred to as Supplemental Environmental Projects (SEPs). SEPs favored by regulatory agencies include pollution prevention projects and proposals that involve public education or increased public awareness. States typically establish a detailed set of requirements governing what constitutes an acceptable SEP. SEPs can be used by facilities to reduce the sting of civil penalty actions by diverting at least some of the penalty payment to a worthwhile environmental project.

29.6 FEDERAL CRIMINAL PENALTIES

The nation's environmental laws govern a broad range of business behavior. Most of these laws contain criminal liability provisions that can be used to prosecute individuals and organizations for "knowing" or "willful" violations of environmental laws. During the past decade, the traditional methods of punishing serious violations of the federal environmental laws have changed dramatically. Notices of violations and administrative enforcement actions have been replaced by criminal complaints and indictments. Most prosecutions are not directed toward the small "midnight dumpers" of hazardous waste but, rather, toward established, legitimate businesses. Indeed, in examining the list of convicted criminal defendants in recent environmental cases, you will find a number of Fortune 500 companies. Corporate officers and managers have also been convicted of environmental crimes and should be advised of their potential legal exposure by the facility's environmental manager and legal counsel.

29.6.1 Criminal Provisions of Environmental Laws

The federal Clean Air Act, the Clean Water Act, the Resource Conservation and Recovery Act, the Toxic Substances Control Act, the Safe Drinking Water Act, the Emergency Planning and Community Right-to-Know Act,

Table 29.1 Civil enforcement actions

Civil Suit/Penalties	Administrative Orders/ Penalties	Field Citation	Other
	Air		
Injunctive relief	Up to $25,000/day, with total penalty limits	Up to $5,000 for minor violations	Compliance orders
Up to $25,000/day per violation			Administrative orders
	Water		
Injunctive relief	Class I: $10,000/day per violation with $25,000 maximum Class II: $10,000/day per violation with $125,000 maximum	—	Compliance orders
Up to $25,000/day per violation			
	RCRA		
Injunctive relief Up to $25,000/day per violation	Up to $25,000/day per violation Revoke/suspend permits	—	Compliance orders Administrative restraining orders (emergency enforcement)
	Additional regulatory information (e.g., monitoring) with added penalities up to $5,000/day for noncompliance		
	EPCRA		
Injunctive relief	Class I: up to $25,000 per violation Class II: $25,000/day per initial violation	—	Compliance orders
Up to $25,000/day per initial violation			
	TSCA		
Injunctive relief	Up to $25,000/day per violation	—	—

and the Superfund Law all provide for criminal penalties for individuals or companies who knowingly or willfully violate the requirements of these laws. The criminal sanctions established by several key environmental statutes are summarized in Table 29.2.

This table provides only a very general summary of the criminal provisions of these laws. For any specific questions regarding applicable criminal penalties, please consult the relevant statutes and regulations.

Repeat criminal offenders are subject to potentially longer prison terms and higher fines. Congress routinely stiffens the criminal penalty provisions of environmental statutes when they are reauthorized or rewritten. Several states have also passed their own statutes, which contain criminal liability provisions similar to those in the federal law.

29.6.2 Prosecution of Environmental Crimes

Traditionally, criminal prosecution for environmental violations was possible only if an individual knowingly or willfully violated the applicable legal requirements. However, this traditional approach to liability for environmental crimes is rapidly changing. For example, in its 1990 Amendments to the Clean Air Act, Congress created a new crime of ''negligent endangerment.'' The Act now imposes a fine of up to $100,000 or a prison term of up to one year, or both, on any person who negligently releases any hazardous air pollutant or extremely hazardous substance that places another person in imminent danger of death or serious bodily injury. This provision of the law clearly demonstrates the intent of Congress to expand liability for environmental crimes to persons who engage in negligent behavior that has a significant adverse impact on human health. It is likely that Congress will add similar liability provisions to future amendments to environmental laws.

Prosecutors are also attempting to expand the number of individuals potentially responsible for environmental crimes through the use of theories such as the ''responsible corporate officer'' doctrine. Under this doctrine, a responsible corporate official can be found guilty of an environmental crime if he or she had responsibility within an organization to either prevent or promptly correct an environmental violation but failed to act. The government believes that criminal liability should be imposed on such corporate officials even if they did not themselves commit the acts underlying the alleged criminal violations. Although courts have split as to whether the doctrine can be used to establish the mental state necessary to commit an environmental crime, it and other broad legal theories of liability invariably will be considered by the prosecution in evaluating an environmental criminal case.

To encourage the prosecution of environmental crimes, Congress, in the

Table 29.2 Criminal penalty actions

Air (first violations)

	Violation of Emission Restrictions	Failure to pay Fees	Negligent Endangerment	Knowing Endangerment
Prison term, up to	5 years	1 year	1 year	15 years
Fine individual, up to	$250,000	$150,000	$100,000	$250,000
Fine corporate, up to	$500,000	$500,000	$200,000	$1,000,000

Water

	Negligent Violation of Permit or Standards (first violation)	False Statements (first violation)	Knowing Violation of Permit or Standards	Knowing Violations Causing Imminent Danger
Prison term, up to	1 year	2 years	3 years	15 years
Fine, individual, up to	$25,000/day per violation	$10,000	$50,000/day per viola-tion	$250,000
Fine, corporate, up to	N/A	N/A	N/A	$1,000,000

RCRA

	Knowingly Treating, Storing, or Transporting Hazardous Waste Without a Permit (first violation)	False Statements	Knowing Endangerment
Prison term, up to	5 years	2 years	15 years
Fine, individual, up to	$50,000/da	$50,000	$250,000
Fine, corporate, up to			$1,000,000

EPCRA

	Knowing Failure to Notify of a Release of a Hazardous Substance (first violation)
Prison term, up to	2 years
Fine, up to	$25,000

TSCA

	Knowing or Willful Violation
Prison term, up to	1 year
Fine, up to	$25,000 per day

Pollution Prosecution Act of 1990, required EPA to hire and train a significant number of additional criminal investigators. In response, the Environmental Crime Unit of the Department of Justice increased its staff of prosecutors to handle a greater number of environmental criminal prosecutions.

29.6.3 Avoiding Criminal Liability for Violations of Environmental Laws

In light of the current emphasis on enforcing environmental laws through criminal prosecution, the best way of avoiding liability for environmental crimes is to implement a strong environmental compliance program. To ensure compliance, the environmental management function of a facility should be properly organized, adequately staffed, and fully and openly supported by the facility's top management team.

A company's failure to properly manage its environmental matters can easily result in the trauma of criminal prosecution directed toward both the company and responsible individuals. It is clear that the prosecution of environmental crimes will remain a powerful political tool into the next century and that the threat of prosecution for serious environmental crimes is indeed real.

29.7 CITIZEN SUITS

All of the major environmental statutes authorize citizen suits. Citizen suits allow interested parties to sue agencies in order to compel them to perform the duties required by the environmental statutes. More important, citizen suit provisions also allow individuals and organizations to serve as "private attorneys general" and bring legal actions, after appropriate notice, directly against alleged violators to force compliance. Citizen suits are especially attractive to environmental groups inasmuch as the environmental statutes provide for the award of attorneys' fees to parties bringing these suits.

29.7.1 Statutory Authority

Provisions authorizing citizen suits are found in the Clean Air Act, the Clean Water Act, RCRA, Superfund, the Community Right-to-Know Act, the Safe Drinking Water Act (SDWA), and various lesser-known environmental laws. Among the major federal environmental statutes, only the Federal Insecticide, Fungicide, and Rodenticide Act (FIFRA) lacks a citizen suit provision.

In general, these citizen suit provisions provide as follows:

- Citizens may file suit against an alleged polluter or the environmental agency charged with enforcing the law.
- The plaintiff must give the alleged polluter, EPA, and the authorized state adequate notice of the threatened lawsuit. The statutes generally require that this notice must be issued at least 60 days prior to filing suit. Thus, these notifications are commonly referred to as ''60-day notice letters.'' Under certain emergency circumstances, lawsuits may be filed immediately.
- Citizen suits are usually barred where the environmental agency is already diligently prosecuting a civil or criminal action against the alleged violator. There is a great deal of confusion within the courts as to whether this enforcement action must be filed in a court rather than with an administrative agency for it to be an effective protection against a citizen suit.
- Federal environmental citizen suits typically must be filed in a federal District Court.
- EPA is typically given the right to intervene in a citizen suit.
- Costs and attorneys' fees may be awarded to plaintiffs in citizen suit cases.

There is a large body of reported case law interpreting the nuances of environmental citizen suits. For example, whether a citizen can prevail in a lawsuit filed against a polluting facility for past violations of an environmental requirement appears to depend on which statute was violated and the wording of the citizen suit provision. Moreover, if a polluting facility comes into compliance during the course of the citizen suit and there is no ongoing violation of an environmental requirement, the court may dismiss the case as being moot.

29.7.2 Practical Advice Regarding Citizen Suits

Historically, most of the citizen suit cases filed have related to violations of the Clean Water Act and are tied to violations of effluent limitations in National Pollutant Discharge Elimination System (NPDES) permits that are evidenced by the data contained in the discharge monitoring reports (DMRs) filed with the regulatory agencies. This information can easily be accessed by members of the public through the federal Freedom of Information Act or its state equivalent. These cases are very difficult to defend, in that the violating facility has provided the evidence of noncompliance and there are typically no good legal defenses available. Legal experts predict that as companies begin generating data that they are required to provide to regulatory agencies under the terms of their Title V air operating permits, there will be a dramatic increase in Clean Air Act citizen suits.

The majority of citizen suits are settled prior to trial, especially in instances where the citizen claims are based on evidence of noncompliance provided by the violating facility. Citizen groups are usually well aware of their right to recover legal fees, and settlement of these cases almost always requires the payment of the plaintiff's legal fees. Citizen plaintiffs are also interested in effectively remedying the instances of noncompliance and improving the quality of the local environment. Environmental managers should think creatively when working to settle environmental citizen suits and focus on settlement options that will not only resolve the environmental noncompliance but will also enhance the reputation and public image of the facility in the community.

30

MANAGING BEYOND COMPLIANCE

Maintaining the basic environmental management programs described throughout this book to ensure regulatory compliance is absolutely essential. There also are some very good reasons for facilities to implement management programs that go "beyond compliance." These innovative programs can benefit a facility and its management team by

- Reducing environmental management costs
- Minimizing future liabilities
- Reducing process losses and costs
- Enhancing public image and trust

This chapter specifically explores programs such as pollution prevention, chemical minimization, the International Standards Organization's ISO 14000, Life Cycle Assessments, and various Environmental Protection Agency (EPA) voluntary programs that go beyond the requirements of existing environmental laws. Some of these concepts will likely be included in the environmental laws and regulations of the next century.

30.1 POLLUTION PREVENTION

30.1.1 Overview

Several federal statutory regulatory programs, such as the Pollution Prevention Act of 1990 (PPA) and the Form R requirements of the Community

Right-to-Know Law, address and encourage pollution prevention. Although these regulatory programs promote the concept of pollution prevention, require reports, and bring some formality to the process, pollution prevention is primarily a voluntary program. A successful pollution prevention program needs support from both facility and environmental managers to give it substance.

In general, environmental agencies look at pollution prevention as a hierarchy of goals with the following order of priority:

- Source reduction or minimization
- Recycling
- Energy recovery
- Disposal

Agencies tend to discount end-of-pipe control and waste treatment improvements as pollution prevention achievements, preferring to concentrate on source reduction. Yet, practically speaking, pollution prevention should include any and all improvements leading to reduced environmental impacts. Why discourage an environmental manager from decreasing final emissions or discharges by questioning whether these improvements qualify as pollution prevention?

30.1.2 Pollution Prevention Initiatives

There are numerous opportunities for pollution prevention projects within most industrial facilities. The following are among the more obvious examples:

- *Drum Management.* Switch to recyclable totes or other containers to eliminate the use of 55 gallon drums and improve spill prevention management.
- *Chemical Containment.* Expand your Spill Prevention, Control, and Countermeasure (SPCC) plan to include all chemical storage.
- *Hazardous Substance Reductions.* Replace solvents and other materials that result in the creation of hazardous waste with nonhazardous substitutes.
- *Polychlorinated biphenyl (PCB) Elimination.* Remove PCB-containing transformers and capacitors.
- *Process Loss Prevention.* Examine loss of process material within the facility and identify an annual cost associated with those losses. Determine how to reduce losses of these materials.

- *Solid Waste Disposal.* Segregate solid waste streams to allow for recycling of wood, paper, metal, plastics, and other materials.
- *Underground Storage Tanks.* Eliminate a major source of soil and groundwater contamination by removing underground tanks.
- *Volatile Organic Compounds (VOC) Emission Reductions.* Substitute reduced solvent or water-based paints and inks for VOC-producing chemicals.

30.1.3 Managing Pollution Prevention Programs

Pollution prevention programs typically entail major philosophical and operational changes within a facility, and many people naturally resist such change. These reactions can be mitigated by providing incentives for pollution prevention and recognizing successful initiatives. It also is a good idea to establish pollution prevention goals and deadlines to give the program both a defined purpose and a sense of urgency.

In addition, facilities should consider advising both environmental agencies and the general public of their successful pollution prevention programs. Many award programs have been established by federal and state environmental agencies, industry trade associations, and public organizations for pollution prevention initiatives. Look for opportunities to take advantage of those award programs, as this sort of recognition will reinforce the importance of pollution prevention and motivate employees to search for other improvement opportunities.

30.2 CHEMICAL MANAGEMENT AND MINIMIZATION

30.2.1 Overview

As environmental managers, we all know that proper utilization of chemicals is critical to the success of many industrial processes. Increased knowledge about the safety and environmental impacts of certain chemicals has resulted in a multitude of regulations regarding chemical management.

In today's regulatory world, it is essential to reduce the quantity and number of chemicals on-site to only what is required to operate the business. A chemical management program should apply not only to process chemicals but also to such items as solvents, paints, part cleaners, spray cans, and any other chemicals used at the facility. Environmental management of a facility is easier with fewer chemicals because the risk of spills, improper releases, and record-keeping and reporting errors is significantly reduced. Those are also significant associated economic benefits, program, most facilities that implement such a program reduce both their inventory costs and hazardous waste disposal fees.

30.2.2 Chemical Inventory

Chemical management in an industrial facility requires a method of identifying and quantifying the chemicals found on-site. This basic inventory can be used to achieve compliance with various environmental and safety regulations. Major industrial complexes such as oil refineries and chemical pulp mills may have several hundred or more chemical compounds on-site at a given time. The number of chemicals and various regulatory requirements make manual tracking virtually impossible. Therefore, for most facilities, a computerized chemical inventory process is required.

Many industrial facilities have coordinated the chemical inventory process with their purchasing practices and accounting systems. For instance, purchasing contracts require vendors to provide "material safety data sheets" for each shipment of chemicals before the shipment will be accepted by the facility. Vendors have also become more sophisticated and realize the importance of supplying this information to their customers.

Another potential ally in the chemical inventory automation process is the accounting department. Because purchasing records are normally coordinated with accounting, many accounting systems can be modified to track the inventory of chemical materials on an ongoing basis. The accuracy of this inventory information can be verified through the normal accounting audit process and periodic field reviews of the chemical inventory.

Once a basic chemical inventory system is established, the environmental manager can easily fulfill most reporting requirements. The real environmental management opportunity lies in minimizing the number and amount of chemicals used at the facility.

30.2.3 Chemical Minimization

Establishing a chemical minimization program will reduce the number and quantity of hazardous chemicals at a facility, reduce operating costs, and reduce the facility's long-term risk of legal exposure. Despite these advantages, many facilities have never implemented such a program. The only apparent explanation for this obvious oversight is the lack of education provided to facility management on the benefits of chemical minimization and the perceived loss of workplace freedom to production managers who prefer complete latitude to decide what chemicals are used to operate the facility.

In order to address these organizational issues and successfully implement

Table 30.1 Chemical Minimization Program

GOAL: TO REDUCE ENVIRONMENTAL RISK AND OPERATING COSTS

STEPS TO BE TAKEN:

1. Eliminating outdated chemicals stored on-site
2. Reducing quantities of hazardous chemicals
3. Substituting nonhazardous chemicals where feasible
4. Restricting new or trial chemicals to preapproved applications
5. Educating employees on chemical management policies and overall chemical goals

a chemical minimization program, full support of the facility's top management is required. Anything less can, and most likely will, diminish the probability of success. The goals of a chemical minimization program must also meet the particular needs of your facility. A suggested program goal and a ...t of steps to be taken are presented in Table 30.1.

...xisting Chemical Review

...nical inventory provides the starting point for reviewing ex- ...to identify minimization candidates. Try to use an open- ...nd carefully consider all chemicals used at the facility, ...sential to the plant's production process. For example, ...d (VOC) emissions caused by solvents in printing ...tly in California to meet state regulatory requi- ...on of low-solvent inks. Yet the use of low- ...ckly spread to other areas of the country ...ness as usual'' attitudes and lack of in- ...vent substitutes.

...uld involve a cross-functional team ...n, technical, maintenance, pur- ...he review should identify any ...tal elimination, reductions, ...ical reformulations, and ...aints used at a facility ...or standard or sim- ...be noted at this ...ay lead to pe- ...d by the facility. ...ated from the process ...t of chemicals carefully. ... chemical minimization pro-

Table 30.2 Chemical minimization program outline

1. Establish a baseline chemical inventory and tracking system.
2. Review existing chemicals for each of the following criteria:
   - Essential to the process, product, or operation
   - Potential for substitution (nonhazardous)
   - Current quantity requirements
   - Potential for consolidation (brand, type, color, etc.)
   - Environmental/safety risk vs. need
3. Establish and enforce a central review program for *all* newly proposed chemicals (including research trials).
4. Establish a purchasing policy that allows only *approved* chemical purchases.

gram does not transform the facility from a small-to large-quantity generator of hazardous waste.

Once the current inventory of chemicals has been minimized, controls must be established to limit future chemical purchases. This is typically accomplished by a central chemical control system.

30.2.5 Central Chemical Control System

Many industrial facilities control chemical purchases by requiring pre
proval before any new chemical is purchased. Central chemical control
tems typically designate an individual or group of managers to appro
introduction of any new chemical to the facility. The approval au
should consider creating a chemical review committee to review
chemical changes or trials and advise as to the regulatory and scienti
associated with each chemical under consideration. The preappro
must be strongly policed by facility management, as exceptions to
quickly destroy the value of the program.

An overall outline for a chemical management minimizati
summarized in Table 30.2.

30.3 ISO 14000

30.3.1 Overview

International environmental management systems ar
raise the environmental management of a facility b
pliance. "ISO 14000" is a shorthand phrase that ref
of international environmental management stand
dards are being developed by one of the technic

national Organization for Standardization (ISO). ISO is a federation of delegates from countries around the world that establishes standards designed to facilitate international trade by ensuring that materials, products, processes, and services meet certain minimum standards. Compliance with any standards developed by ISO is voluntary; however, it often happens that a country adopts ISO standards and requires companies that market products in that country to comply with the standards. The ISO 14000 standards are intended to establish an international standard for environmental management. The primary focus of this section is on the ISO standard that is used for developing and certifying environmental management systems—ISO 14001.

30.3.2 ISO 14001

ISO 14001 was finalized in September 1996 and is now being implemented by companies throughout the world. ISO 14001 is designed to improve a company's environmental performance through the creation and implementation of an Environmental Management System (EMS). A companion standard, ISO 14004, provides guidance on developing and implementing an EMS and integrating the EMS into the overall management practices at the facility.

The key elements of ISO 14001 are as follows:

- *Environmental Policy.* An appropriate environmental policy must be documented and communicated to employees and made available to the public. The policy should include a commitment to continuous improvement, pollution prevention, and regulatory compliance. The policy should also provide a framework for establishing environmental objectives and targets.
- *Planning.* The organization should develop a procedure that allows it to identify the environmental aspects and applicable legal requirements relating to its activities, establish and document objectives and targets consistent with its environmental policy, and create a program for meeting those targets and objectives (including the designation of responsible individuals, necessary means, and appropriate time frames).
- *Implementation and Operation.* The organization should devote adequate resources to implement and operate the EMS. The organization should define, document, and clearly communicate environmental management responsibilities. The organization must also identify training needs and provide employees with appropriate training. Internal and external communication regarding the EMS is another important re-

quirement of ISO 14001, and organizations must document these procedures and provide adequate document control. The organization must also establish and maintain procedures to ensure that its operations meet the requirements of the EMS. Emergency preparedness and response planning is another essential EMS requirement.

- *Checking and Corrective Action.* Checking and corrective action procedures must be developed by the organization to ensure regular monitoring and measurement of key process characteristics that have a significant impact on the environment. The organization must develop a documented procedure for evaluating compliance with applicable environmental requirements and correcting any noncompliance. The organization must also establish a system for managing environmental records and conducting periodic audits of the EMS.
- *Management Review.* The organization's top management must periodically review the overall EMS to ensure its suitability, adequacy, and effectiveness in light of changing circumstances.

Once a company has developed an EMS that meets the requirements of ISO 14001, the next question is whether to obtain ISO 14001 certification.

Companies that choose to obtain ISO 14001 certification contract a private registrar certified by the American National Standards Institute/Registration Accreditation Board (ANSI/RAB). The registrar sends a group of auditors to the company's facility to determine whether the company's EMS meets the ISO 14001 standard. This process is commonly referred to as the registration audit.

Companies that have an approved EMS are provided with a certificate that bears the name of the registrar and the accreditation body. The certificates are valid for a finite period of time, usually three years. The registrars conduct surveillance audits during the life of the certificate to ensure that the company continues to meet the ISO 14001 requirements.

30.3.3 Why Implement ISO 14001?

The major advantage of ISO 14001 is that it provides a widely accepted method of designing and implementing an environmental management system. Whether a company needs to ISO certify, its EMS is influenced by several major factors.

First, and probably most important, there likely will be strong market forces pushing companies to ISO 14001 certify. For example, a number of companies have already obtained ISO 14001 certification and are requiring their suppliers to certify. This trend will likely continue into the future.

There may also be strong regulatory incentives to ISO certify environmental management systems. ISO 14000 concepts have already been evaluated by the Environmental Protection Agency (EPA) in connection with such programs as the Common Sense Initiative, Project XL, and the Environmental Leadership Program. Both EPA's audit policy and the Department of Justice guidance on environmental criminal prosecutions have provisions that favor companies with ISO 14000–type management systems. Several state regulatory agencies are considering offering more lenient regulatory standards for companies that ISO 14001 certify.

Finally, many companies will consider ISO 14001 certification to be another way to demonstrate to the public they are good corporate citizens. These pressures alone may be enough to persuade many companies to ISO certify.

Compliance with ISO 14001 or a similar environmental management system standard will likely be required of most companies within the next decade. Environmental managers should understand ISO 14001 and its requirements in order to prepare themselves for the environmental management challenges of the next century.

30.4 EPA INDUSTRY PARTNERSHIPS

30.4.1 Beyond the Command and Control Approach

Industrial facilities are being encouraged to move beyond basic regulatory compliance through numerous voluntary initiative programs developed by EPA. These programs often conflict with the historical command-and-control approach to environmental compliance taken by EPA and mark a subtle shift in the way EPA views environmental regulation. For example, in the early years of the Community Right-to-Know Program, EPA established a voluntary reduction program called the ''EPA 33/50 Program.'' It was aimed at enlisting volunteers to reduce releases of 17 listed toxic chemicals by 50% over five years. Many companies participated and major reductions were achieved, often far exceeding EPA's 50% goal in less than the five-year time frame. EPA, industry, and the environment all benefited from the agency's 33/50 Program.

Other EPA programs are aimed at energy efficiency, which have the indirect benefit of reducing air emissions and greenhouse gases. EPA's Greenlights Program, which encourages the voluntary use of energy-efficient lighting, is an example of one of the agency's energy efficiency programs.

The list of voluntary opportunities to partner with EPA is expanding. As

Table 30.3 EPA industry partnerships

EPA Business Initiatives and Assistance Programs. A listing of EPA Business Initiatives and compliance assistance information.

Design for the Environment. A voluntary program designed to help businesses incorporate environmental considerations into the design and redesign of products, processes, and technical and management systems.

Electronic Commerce/Electronic Data Interchange (EC/EDI). A program working to introduce electronic reporting for all major environmental compliance programs, both for reports submitted directly to EPA and for those submitted to state or local agencies under delegated programs.

Energy Star. Voluntary programs and products designed to promote energy efficiency, reduce pollution, and save money for consumers, organizations, and businesses.

Environmental Accounting Project. A voluntary program to encourage and motivate businesses to understand the full spectrum of their environmental costs and integrate these costs into decision making.

Environmental Technology Initiative. A partnership program to promote improved public health and environmental protection by advancing the development and use of innovative environmental technologies.

Green Lights. A voluntary program that encourages the widespread use of energy-efficient lighting.

Methane Outreach. Voluntary programs designed to promote profitable opportunities for reducing methane and other greenhouse gas emissions in the United States.

Pesticide Environmental Stewardship Program. A voluntary program that forms partnerships with pesticide users to reduce health and environmental risk and implement pollution prevention strategies.

Project XL. A national pilot program that tests innovative ways of achieving better and more cost-effective public health and environmental protection.

Reinvention for Innovative Technologies (ReFIT). An interagency program to improve public health and environmental protection by removing barriers to technology innovation.

Sustainable Industry Project. A program designed to explore, design, and promote industry sector–based approaches to environmental protection.

Waste Wise. A voluntary program that targets the reduction of municipal solid waste that would otherwise end up in a trash dumpster.

outlined in Table 30.3, the current list of programs at EPA's website is quite extensive.

30.4.2 Project XL

An EPA program that is especially worthy of further discussion is Project XL. This program is described by EPA as a national pilot program that tests innovative ways to protect the environment and human health. Is there a better way to meet the objectives of the environmental laws than traditional command and control regulations? If so, EPA may consider the revised strategy under Project XL. EPA has established a goal of implementing 50 pilot XL projects in industry, government organizations, and communities. XL projects are especially attractive to industry when EPA is willing to exercise regulatory flexibility and recognize the particular circumstances of a facility.

EPA asks that XL applicants have a strong, well-established compliance record. Any proposed XL project should include an environmental strategy that:

- Improves final results
- Reduces costs and paperwork under current regulations
- Is supported by other stakeholders
- Is innovative and pollution-prevention oriented
- Is transferable to others if successful
- Does not transfer risk to others

The successful applicant will have to work closely with the local community, state agencies, tribal governments, and other affected parties to gain support for the project. If consensus among these groups is reached, EPA will take a much greater interest in reviewing an XL proposal and working toward approval.

30.5 LIFE CYCLE ASSESSMENTS

An environmental management concept of increasing importance is the Life Cycle Assessment (LCA). LCA takes environmental impact analysis to a new level and will likely be an important part of environmental management throughout the next century. LCA evaluates the impacts of a facility's product or process from "cradle to grave." The Society for Environmental Toxicology and Chemistry (SETAC) is an organization that has been instrumental in developing the ground rules for LCA. SETAC defines LCA as follows:

> Life cycle assessment is an objective process to evaluate the environmental burdens associated with a product, process, or activity by identifying and quantifying energy and materials used and wastes released to the environment, to assess the impact of those energy and material uses and releases to the environment and to evaluate and implement opportunities to affect environmental improvements. The assessment includes the entire life cycle of the product, process, or activity, encompassing extracting and processing raw materials;

manufacturing, transportation, and distribution; use, re-use, maintenance, recycling, and final disposal.*

The LCA approach has gained major support in the European community. If a company exports to Europe, an LCA evaluation of its product line may be required before the product is allowed to enter the European market.

The concept is noble. However, the rules for the performance of a product LCA are still subject to debate. SETAC has published *A Conceptual Framework for Product Life-Cycle Impact Assessment* which gives guidance on how to perform such an assessment. EPA has also published a document entitled *The Use of Life Cycle Assessment in Environmental Labeling.* Both documents are useful in learning more about how the LCA process may apply to your specific product and facility.

SETAC describes a product life cycle assessment as illustrated in Figure 30.1.

The diagram includes four areas to consider in the LCA process:

- *Goal Definition and Scoping.* This area includes defining the purpose of the study, the scope of the study and the quality assurance procedure to be followed. In addition, units of performance measure are identified, such as the amount of a chemical required to perform a process.
- *Inventory Analysis.* The technical analysis of data available where quantification of emissions, energy requirements, or raw material requirements are identified.
- *Impact Assessment.* Impacts are analyzed systematically to evaluate potential environmental impacts on humans, ecosystems, and natural resources. Although a less-is-best approach was taken originally, more consideration is now being given to trade-offs, renewable resources, and holistic evaluations.
- *Improvement Assessment.* The qualitative evaluation of impacts to determine whether improvements can be made to a product or management process over the life of the product or process.

The LCA process involves a great deal of subjectivity, and the results can be influenced by the goal definition and scoping process. For example, in comparing the life cycles of plastic and paper bags, do you start with the manufacturing process or go back to the source of the raw material? In evaluating the raw material, what has a greater environmental impact, harvesting a forest or pumping oil from a petroleum deposit?

*Reprinted with permission from *Guidelines for Life-Cycle Assessment: A ''Code of Practice,''* Consoli, F., et al., Editors, Copyright Society of Environmental Toxicology and Chemistry (SETAC), Pensacola, FL, 1993.

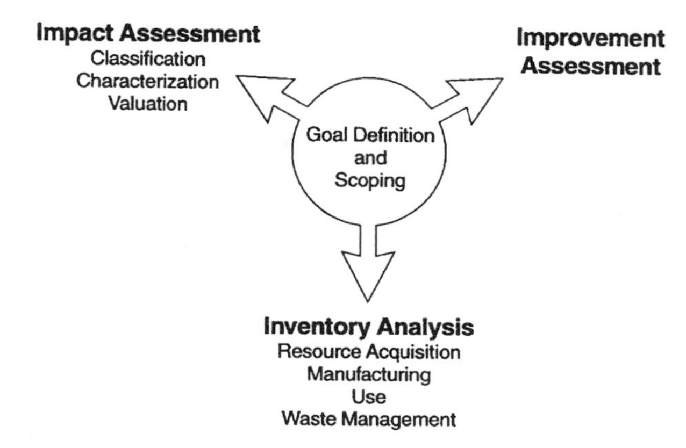

Figure 30.1 Product life cycle assessment (SETAC, 1993). Reprinted with permission from *Guidelines for Life-Cycle Assessment: A "Code of Practice,"* Consoli, F., et al., Editors, Copyright Society of Environmental Toxicology and Chemistry (SETAC), Pensacola, FL, 1993.

A simpler form of the LCA process used in Germany is called the Blue Angel Program. This program is aimed at labeling products to help consumers identify the environmental impacts associated with production, use, and disposal of those products. Figure 30.2 shows the evaluation matrix used in the program.

The Germans have used this approach to evaluate a variety of products. The evaluation conducted for copying machines and printers led some companies to institute programs to allow the return of toner cartridges as an improved method of waste disposal.

Overall, the LCA process is still developing. It is an area of emerging importance for environmental managers and another example of the growing overlap between product marketing and environmental issues.

	Production	Use	Disposal
Hazardous Substances			
Emissions			
Air			
Water			
Soil			
Noise			
Waste Minimization			
Resource Conservation			
Fitness for Use			
Safety			

Figure 30.2 German "Blue Angel" evaluation chart

APPENDIX A
ACRONYMS

AAP: Asbestos Action Program

A&C: abatement and control

ACE: any credible evidence

ACM: Asbestos-containing materials

ACQR: Air Quality Control Region

ACTS: Asbestos Contractor System

AHERA: Asbestos Hazard Emergency Response Act

AL: acceptable level

ALARA: as low as reasonably achievable

ANSI: American National Standards

AOP: Air Operating Permit

APCD: Air Pollution Control District

API: American Petroleum Institute

AQA: Air Quality Act

AQCP: Air Quality Control Program

AQCR: Air Quality Control Region

AQMD: Air Quality Management District

ARAR: Applicable or Relevant and Appropriate Requirements

ASTM: American Society for Testing and Materials

BACT: Best Available Control Technology

BADT: Best Available Demonstrated Technology

BAF: Bioaccumulation Factor

BART: Best Available Retrofit Technology

BAT: Best Available Technology, Best Available Control Technology Economically Achievable

BCT: Best Conventional Pollutant Control Technology, Best Control Technology

BDAT: Best Demonstrated Achievable Technology

BDT: Best Demonstrated Technology

BEJ: Best Engineering Judgment, Best Expert Judgment

BEN: EPA Economic Benefit Model for Penalties

BMP: Best Management Practices

BOD: Biochemical Oxygen Demand, Biological Oxygen Demand

BPT: Best Practice Technology, Best Practicable Treatment

Btu: British Thermal Unit

CA: Citizen Act, Competition Advocate, Cooperative Agreements, Corrective Action

CAA: Clean Air Act

CAAA: Clean Air Act Amendments

CAM: Compliance Assurance Monitoring

CAO: Corrective Action Order

CAS: Chemical Abstract Service Number

CBI: Confidential business information

CBOD: carbonaceous biochemical oxygen demand

CEM: continuous emission monitoring

CEQ: Council on Environmental Quality

CERCLA: Comprehensive Environmental Response, Compensation, and Liability Act of 1980 (Superfund)

CERCLIS: CERCLA Information System

CFC: chlorofluorocarbon

cfm: cubic feet per minute

CFR: Code of Federal Regulations

CH4: methane

CIH: Certified Industrial Hygienists

ClO_2: chlorine dioxide

CIS: Chemical Information System, Contracts Information System

Cl: Chlorine

CMA: Chemical Manufacturers Association

CMB: chemical mass balance

CO: carbon monoxide

CO_2: carbon dioxide

COD: chemical oxygen demand

CPF: Carcinogenic Potency Factor

CRTK: Community Right-to-Know Act (see EPCRA and SARA)

CWA: Clean Water Act

DMR: Discharge Monitoring Report

DO: dissolved oxygen

DOE: Department of Energy, Department of the Environment, Department of Ecology

DOJ: Department of Justice

DOT: Department of Transportation

DPA: Deepwater Ports Act

DWEL: drinking water equivalent level

DWS: Drinking Water Standard

EA: Endangerment Assessment, Enforcement Agreement, Environmental Action, Environmental Assessment, Environmental Audit

ECRA: Environmental Cleanup Responsibility Act

ECU: Environmental Crimes Unit

EDF: Environmental Defense Fund

EER: Excess Emission Report

EF: emission factor

EHS: extremely hazardous substance

EIS: Environmental Impact Statement, Environmental Inventory System

EL: exposure level

ELI: Environmental Law Institute

ELR: *Environmental Law Reporter*

EMS: Environmental Management Standards, Environmental Management System

EOP: end of pipe

EPA: Environmental Protection Agency

EPACT: Environmental Policy Act

EPCRA: Emergency Planning and Community Right-to-Know Act

ERA: ecological risk assessment, Economic Regulatory Agency

ERL: Environmental Research Laboratory

ERT: emergency response team

ES: enforcement strategy

ESA: Endangered Species Act

ESH: environmental safety and health

ESP: electrostatic precipitators

ET: emissions trading
FE: fugitive emissions
FGD: flue-gas desulfurization
FID: flame ionization detector
FIFRA: Federal Insecticide, Fungicide, and Rodenticide Act
FIM: friable insulation material
FINDS: Facility Index System
FIP: Federal Implementation Plan
FMP: Facility Management Plan, Financial Management Plan
FOE: Friends of the Earth
FPA: Federal Pesticide Act
FS: feasibility study
FWPCA: Federal Water Pollution Control Act
FWS: U.S Fish and Wildlife Service
GC/MS: gas chromatograph/mass spectrograph
GEMI: Global Environmental Management Initiative
GEMS: Global Environmental Monitoring System, Graphical Exposure Modeling System
GLNPO: Great Lakes National Program Office
gpm: gallons per minute
GVW: gross vehicle weight
GWM: groundwater monitoring
GWP: global warming potential
H: hydrogen atom
HAP: hazardous air pollutant
HAZMAT: hazardous materials
HCFC: hydrochlorofluorocarbon
HEPA: high-efficiency particulate air
HFC: hydrofluorocarbon
HHV: higher heating value
HI: Hazardous Index
HMR: Hazardous Materials Regulations
HMTA: Hazardous Materials Transportation Act
HMTR: Hazardous Materials Transportation Regulations
hp: horsepower
HQ: Hazard Quotient

HRA: Health Risk Assessment
HRS: Hazard ranking system
HSWA: Hazardous and Solid Waste Amendments of 1984
HW: hazardous waste
HWIR: Hazardous Waste Identification Rule
HWM: Hazardous Waste Management
IAQ: indoor air quality
ICS: Individual Control Strategy
IEA: Institute for Environmental Auditing
ISC: Industrial Source Complex
ISO: International Standards Organization
LAER: lowest achievable emission rate
LCA: Life Cycle Assessment
LDR: Land Disposal Restrictions
LDS: leak detection system
LEP: Laboratory Evaluation Program
LEPC: local emergency planning committee
LOAEL: lowest observable adverse effect level
LOEC: lowest observed effect concentration
LUST: leaking underground storage tank
MACT: Maximum Achievable Control Technology
MCL: maximum contaminant level
MDL: method detection limit
MEI: maximally (or most) exposed individual
mg/l: milligrams per liter = parts per million
MPI: maximum permitted intake
MSDS: Material Safety Data Sheet
MSW: municipal solid waste
MTCA: Model Toxics Control Act
MUTA: mutagenicity
NAAQS: National Ambient Air Quality Standards
NAS: National Academy of Sciences
NCP: National Contingency Plan
NEIC: National Enforcement Investigation Center
NEPA: National Environmental Policy Act
NESHAP: National Emission Standards for Hazardous Air Pollutants

NIEH: National Institutes for Environmental Health

NIMBY: not-in-my-backyard

NIOSH: National Institute for Occupational Safety and Health

NMFS: National Marine Fisheries Service

NO: nitric oxide

NO_2: nitrogen dioxide

NO_x: nitrogen oxides

N_2O: Nitrous Oxide

NOAA: National Oceanic and Atmospheric Administration

NOAEL: No Observable Adverse Effect Level

NOC: Notice of Construction

NOEC: No Observable Effects Concentration

NOEL: No Observable Effect Level

NOI: Notice of Intent to Construct

NON: Notice of Noncompliance

NOV: Notice of Violation

NPDES: National Pollutant Discharge Elimination System

NPL: National Priorities List

NRC: Nuclear Regulatory Commission, National Response Center

NRDC: Natural Resources Defense Council

NSDWR: National Secondary Drinking Water Regulations

NSPS: New Source Performance Standards

NSR: New Source Review

NTP: National Toxicology Program

O: Oxygen

O_2: oxygen molecules

O_3: ozone

O&M: Operations and maintenance

ODC: ozone-depleting chemical

ODP: ozone-depleting potential

ODS: ozone-depleting substances

OHMT: Office of Hazardous Materials Transportation

OPA: Oil Pollution Act

OSHA: Occupational, Safety, and Health Act: Occupational Safety and Health Administration

OTR: ozone transport region

P2: Pollution Prevention

Pb: Lead

PCB: polychlorinated biphenyl

PCDDs: polychlorinated dibenzo-p-dioxins

PCP: pentachlorophenol

PDWS: Primary Drinking Water Standard

PEL: permissible exposure level

PM: particulate matter

$PM_{2.5}$: Particulate matter less than or equal to 2.5 microns (diameter)

PM_{10}: Particulate Matter less than or equal to 10 microns (diameter)

PMN: Premanufacture Notification

POC: point of compliance

POE: point of exposure

POI: point of interception

POM: particulate organic matter, polycyclic organic matter

POTW: publicly owned treatment works

PPA: Pollution Prevention Act, Pollution Prosecution Act

ppb: parts per billion

ppt: parts per trillion

PRA: Paperwork Reduction Act, planned regulatory action

PRP: potentially responsible party

PS: point source

PSD: prevention of significant deterioration

PSL: Priority Substances List

PSAM: Point Source Ambient Monitoring

PSM: process safety management, point source monitoring

PTC: Permit to Construct

PUC: Public Utility Commission

PWS: public water supply/system

PWSS: public water supply system

QAC: Quality Assurance Coordinator

QA/QC: Quality Assurance/Quality Control

QL: quantification limit

RA: reasonable alternative, regulatory alternatives, regulatory analysis, remedial action, resource allocation, risk analysis, risk assessment

RAB: Registration Accreditation Board

RACM: regulated asbestos-containing material, reasonably available control measures
RACT: Reasonably Available Control Technology
RCRA: Resource Conservation and Recovery Act
R&D: research and development
RDF: refuse-derived fuel
RFD: reference doses
RFI: RCRA facility investigation
RI: Remedial investigation
RI/FS: Remedial investigation/feasibility study
RMP: Risk Management Plan, Risk management plan
RMW: regulated medical waste
ROD: Record of decision
RQ: reportable quantity
RTDM: Rough Terrain Diffusion Model
RTM: Regional Transport Model
RTP: Research Triangle Park
SAP: Scientific Advisory Panel
SAR: Safety Analysis Review
SARA: Superfund Amendments and Reauthorization Act of 1986
SC: Sierra Club
SCLDF: Sierra Club Legal Defense Fund
SDWA: Safe Drinking Water Act
SEP: supplemental environmental project
SEPA: State Environmental Policy Act
SERC: State Emergency Response Commission
SETAC: Society for Environmental Toxicology and Chemistry
SIC: Standard Industrial Classification
SIP: State Implementation Plan
SO_2: sulfur dioxide
SOC: synthetic organic chemicals
SOCMI: synthetic organic chemicals manufacturing industry
SOTDAT: source test data
SPCC: Spill Prevention, Control, and Countermeasure (Plan)
SPS: state permit system
STEL: short-term exposure limit

SWDA: Solid Waste Disposal Act
TAMS: Toxic Air Monitoring System
TAP: Technical Assistance Program
TCDD: dioxin (tetrachlorodibenzo-p-dioxin)
TCF: totally chlorine free
TCLP: Total Concentrate Leachate Procedure, Toxicity Characteristic Leachate Procedure
TCRI: Toxic Chemical Release Inventory
TDS: Total dissolved solids
TMDL: Total maximum daily load
TPQ: threshold planning quantity
tons/year: tons per year
TQM: Total Quality Management
TRI: Toxics Release Inventory
TSCA: Toxic Substances Control Act
TSD: treatment, storage, and disposal, technical support document
TSDF: treatment, storage, and disposal facilities
TSP: total suspended particulates
TSS: Total suspended solids
TWA: time-weighted average
UIC: underground injection control
USC: United States Code
UST: underground storage tank
VOC: volatile organic compounds
VP: vapor pressure
WET: Whole Effluent Toxicity Test
WQA: Water Quality Act of 1987
WQS: Water Quality Standard
ZRL: zero risk level

APPENDIX B

GLOSSARY

Abandoned Well: A well whose use has been permanently discontinued or that is in a state of such disrepair that it cannot be used for its intended purpose.

Abatement: Reducing the degree or intensity of, or eliminating, pollution.

Absorbed Dose: In exposure assessment, the amount of a substance that penetrates an exposed organism's absorption barriers (e.g., skin, lung tissue, gastrointestinal tract) through physical or biological processes. The term is synonymous with *internal dose*.

Absorption: The uptake of water, other fluids, or dissolved chemicals by a cell or an organism (as tree roots absorb dissolved nutrients in soil).

Action levels: In the Superfund program, the existence of a contaminant concentration in the environment high enough to warrant action or trigger a response under SARA and the National Oil and Hazardous Substances Contingency Plan.

Activated sludge: Product that results when primary effluent is mixed with bacteria-laden sludge and then agitated and aerated to promote biological treatment, speeding the breakdown of organic matter in raw sewage undergoing secondary waste treatment.

Acute exposure: A single exposure to a toxic substance which may result in severe biological harm or death. Acute exposures are usually characterized as lasting no longer than a day, as compared to longer, continuing exposure over a period of time.

Acutely hazardous wastes: Toxic waste materials that, in small doses or exposures, are capable of causing death or significantly contributing to irreversible and/or incapacitating illness.

Acute Toxicity: The ability of a substance to cause severe biological harm or death soon after a single exposure or dose. Also, any poisonous effect resulting from a single short-term exposure to a toxic substance.

Administrative Order: A legal document signed by EPA directing an individual, business, or other entity to take corrective action or refrain from an activity. It describes the violations and actions to be taken and can be enforced in court. Such orders may be issued, for example, as a result of an administrative complaint whereby the respondent is ordered to pay a penalty for violations of a statute.

Administrative Procedures Act: A law that spells out procedures and requirements related to agency actions and the promulgation of regulations.

Adsorption: Removal of a pollutant from air or water by collecting the pollutant on the surface of a solid material; (e.g., an advanced method of treating waste in which activated carbon removes organic matter from wastewater).

Aerated lagoon: A holding and/or treatment pond that speeds up the natural process of biological decomposition of organic waste by stimulating the growth and activity of bacteria that degrade organic waste.

Aerosol: Small droplets or particles suspended in air or another gaseous environment.

Air pollutant: Any substance in air that can, in high enough concentration, harm humans, other animals, vegetation, or material. Pollutants include almost any natural or artificial composition of matter capable of being airborne. They may be in the form of solid particles, liquid droplets, gases, or in combination thereof.

Air pollution: The presence of contaminants or pollutant substances in the air that interfere with human health or welfare, or produce other harmful environmental effects.

Air pollution control device: Mechanism or equipment that cleans emissions generated by a source (e.g., an incinerator, industrial smokestack, or automobile exhaust system) by removing pollutants that would otherwise be released to the atmosphere.

Air shed: The air impacted by an emission source in a given geographical area and confined by physical boundaries such as hills or alleys.

Air strippers: Devices that are used for removing volatile products such as petroleum constituents in groundwater remediation work. The process consists of discharging a waste stream at the top of a tower while simultaneously blowing air upward through the water stream. The passage of the air strips the volatile contaminants within the water stream.

Air toxics: Air pollutants for which national ambient air quality standards (NAAQS) do not exist (i.e., excluding ozone, carbon monoxide, PM_{10}, sulfur dioxide, nitrogen oxide) that may reasonably be anticipated to cause cancer; respiratory, cardiovascular, or developmental effects; reproductive dysfunctions, neurological disorders, heritable gene mutations, or other serious or irreversible chronic or acute health effects in humans.

Alkaline: The condition of water or soil that contains a sufficient amount of alkali substance to raise the pH above 7.0.

Ambient air: Any unconfined portion of the atmosphere: open air, surrounding air.

Aquatic ecosystems: Ecosystems that may be primarily fresh water (lakes, ponds, streams, rivers) or marine (ocean, seacoast, estuaries).

Aquifer: An underground geological formation, or group of formations, containing water. Aquifers are sources of groundwater for wells and springs.

Aquitard: Geologic formation that may contain groundwater but is not capable of transmitting significant quantities of it under normal hydraulic gradients. May function as a confining bed.

Aromatics: A type of hydrocarbon, such as benzene or toluene, with a specific type of ring structure. Aromatics are sometimes added to gasoline to increase octane. Some aromatics are toxic.

Article: A manufactured item that is formed to a specific shape or design during manufacture, that has end use functions, and that does not release or otherwise result in exposure to a regulated substance under normal conditions of process and use.

Asbestos: A mineral fiber that can pollute air or water and cause cancer or asbestosis when inhaled. EPA has banned or severely restricted its use in manufacturing and construction.

Background level: (1) The concentration of a substance in an environmental media (air, water, or soil) that occurs naturally or is not the result of human activities. (2) In exposure assessment the concentration of a substance in a defined control area, during a fixed period of time before, during, or after a data-gathering operation.

BACT—best available control technology: An emission limitation based on the maximum degree of emission reduction (considering energy, environmental, and economic impacts) achievable through application of production processes and available methods, systems, and techniques.

Baghouse filter: Large fabric bag, usually made of glass fibers, used to eliminate intermediate and large particles. The device operates like the bag of an electric vacuum cleaner, passing the air and smaller particles while entrapping the larger ones.

Banking: A system for recording qualified air emission reductions for later use in bubble, offset, or netting transactions.

Bar screen: In wastewater treatment, a device used to remove large solids.

BEN: EPA's computer model for analyzing a violator's economic gain from not complying with the law.

Benefit-cost analysis: An economic method for assessing the benefits and costs of achieving alternative health-based standards at given levels of health protection.

Best demonstrated available technology: (BDAT) As identified by EPA, the most effective commercially available means of treating specific types of hazardous waste. BDATs may change with advances in treatment technologies.

Best management practice (BMP): Methods that have been determined to be the most effective, practical means of preventing or reducing pollution from nonpoint sources.

Bioaccumulation: The extent to which chemicals accumulate in living organisms such as fish or other animals.

Bioassay: A test to determine the relative strength of a substance by comparing its effect on a test organism with that of a standard preparation.

Biodegradable: Capable of decomposing under natural conditions.

Biological oxygen demand (BOD): A measure of the amount of oxygen consumed in the biological processes that break down organic matter in water. The greater the BOD, the greater the degree of oxygen required. The test result is also dependent on time duration; BOD tests are normally run on a five-day basis (BOD*5*).

Biological treatment: A treatment technology that uses bacteria to consume organic waste.

Bioremediation: Use of living organisms to clean up oil spills or remove other pollutants from soil, water, or wastewater; use of organisms such as nonharmful insects to remove agricultural pests or counteract diseases of trees, plants, and garden soil.

Bioremediation technologies: Methods that enhance biodegradation of contaminants through the stimulation of indigenous soil and groundwater microbial populations or the addition of natural microbial species.

British thermal unit (Btu): Unit of heat energy equal to the amount of heat required to raise the temperature of 1lb of water by one degree Fahrenheit at sea level.

Carbon monoxide (CO): A colorless, odorless, poisonous gas produced by incomplete fossil fuel combustion.

Carcinogens: Substances that may interact with genetic material and stimulate uncontrolled cell proliferation through one or more mechanisms, causing cancer.

CAS registration number: A number assigned by the Chemical Abstract Service to identify a chemical.

Chemical oxygen demand (COD): A measure of the oxygen required to oxidize all compounds, both organic and inorganic, in water.

Chemical treatment: Any one of a variety of technologies that use chemicals or a variety of chemical processes to treat waste.

Chlorination: The application of chlorine to drinking water, sewage, or industrial waste to disinfect or to oxidize undesirable compounds.

Chlorofluorocarbons: Synthetic hydrocarbon compounds containing chlorine and fluorine. They have properties that make them desirable as refrigerants, solvents, and in other industrial uses.

Clarifier: A tank in which solids settle to the bottom and are subsequently removed as sludge.

Class I area: Under the Clean Air Act, a Class I area is one in which visibility is protected more stringently than under the National Ambient Air Quality Standards; includes national parks, wilderness areas, monuments, and other areas of special national and cultural significance.

Clean Air Act (CAA): The federal law that establishes goals for emission reductions, ambient air quality improvements, and air permit programs for specified pollutants.

Clean Water Act (CWA): A federal law that limits the discharge of pollutants into navigable waters with the goal of making the nation's waters fishable and swimmable.

Coliform bacteria: Bacteria associated with human or animal wastes and potentially capable of causing disease.

Compliance: Meeting the intent of a request, command, regulation, permit condition, etc.

Compliance schedule: Specific timetable by which a facility must bring itself into compliance with an applicable requirement. Such schedules may be set forth in regulations, special judicial orders issued by a court, or administrative orders issued by an environmental agency.

Compliance Citations: Official citations by federal or state environmental agencies.

Conditionally exempt small-quantity generator: A generator that does not generate more than about 220 lb or 25 gal (100 kg) of hazardous waste and no more than 2.2 lb (1 kg) of acute hazardous waste in a month. A conditionally exempt small-quantity generator is not subject to the same hazardous waste management and permitting requirements as generators above that level.

Confidential business information (CBI): Material that contains trade secrets or commercial or financial information that has been claimed as confidential by its source (e.g., a pesticide or new chemical formulation registrant). EPA has special procedures for handling such information.

Consent decree: A legal document, approved by a judge, that formalizes an agreement reached between EPA or another agency and a facility to resolve alleged legal violations.

Contingency plan: A plan, readily available on-site that contains provisions for environmental response in case hazardous materials spill on-site. Any plan for advance preparation.

Criteria pollutants: The 1970 Amendments to the Clean Air Act required EPA to set National Ambient Air Quality Standards for certain ''criteria pollutants'' known to be hazardous to human health. EPA has identified and set standards to protect human health and welfare for six pollutants: ozone, carbon monoxide, total suspended particulates, sulfur dioxide, lead, and nitrogen oxide.

Decommissioning: The process of retiring something from service.

Degradation: The transformation of a contaminant to other forms through decay, biodegradation, or other process.

Deposition: The process by which particles and reactive gases are deposited on the earth's surface from the atmosphere.

Dermal absorption/penetration: Process by which a chemical penetrates the skin and enters the body as an internal dose.

Dermal exposure: Contact between a chemical and the skin.

Dilution ratio: The relationship between the volume of water in a stream and the volume of incoming water; affects the ability of the stream to assimilate waste.

Dioxin: Any of a family of compounds known chemically as dibenzo-p-dioxins. Concern about these compounds arises from their potential toxicity as contaminants in commercial products. Tests on laboratory animals indicate that these are among the more toxic anthropogenic (man-made) compounds.

Dispersion: Spreading of a contaminant source as it flows through an aquifer or air shed.

Disposal: To discharge, deposit, inject, dump, spill, leak, or place wastes into or on land or waters.

Disposal facilities: Repositories for solid waste, including landfills and combustors intended for permanent containment or destruction of waste materials. Excludes transfer stations and composting facilities.

Dissolved oxygen (DO): The oxygen freely available in water, vital to fish and other aquatic life and for the prevention of odors. The DO level is considered an important indicator of a water body's ability to support desirable aquatic life. Secondary and advanced waste treatment are generally designed to ensure adequate DO in waste-receiving waters.

Dissolved solids: Disintegrated organic and inorganic material in water. Excessive amounts make water unfit to drink or use in industrial processes.

Dosage/dose: (1) The actual quantity of a chemical administered to an organism or to which it is exposed. (2) The amount of a substance that reaches a specific tissue (e.g., the liver). (3) The amount of a substance available for interaction with metabolic processes after crossing the outer boundary of an organism.

Dose response: Shift in toxicological responses of an individual (such as alterations in severity) or populations (such as alterations in incidence) that are related to changes in the dose of any given substance.

Dose response curve: Graphical representation of the relationship between the dose of a stressor and the biological response thereto.

Dredging: Removal of mud from the bottom of water bodies. This can disturb the ecosystem and causes silting that kills aquatic life. Dredging of contaminated muds can expose biota to heavy metals and other toxics.

Ecological risk assessment: The application of a formal framework, analytical process, or model to estimate the effects of human action(s) on nature and to interpret the significance of those effects in light of the uncertainties identified in each component of the assessment process. Such an analysis includes initial hazard identification, exposure and dose-response assessments, and risk characterization.

Ecology: The relationship of living things to one another and their environment, or the study of such relationships.

Ecosystem: The interacting system of a biological community and its non-living environmental surroundings.

Ecosystem structure: Attributes related to the instantaneous physical state of an ecosystem; examples include species population density, species richness or evenness, and standing crop biomass.

Effluent: Wastewater—treated or untreated—that flows out of a treatment plant, sewer, or industrial outfall. Generally refers to wastes discharged into surface waters.

Effluent guidelines: Technical EPA documents that set effluent limitations for given industries and pollutants.

Emission: A discharge of a pollutant to the air.

Emission factor: The relationship between the amount of pollution produced and the amount of raw material processed. For example, an emission factor for a blast furnace making iron would be the number of pounds of particulates per ton of raw materials.

Endangered species: Animals, birds, fish, plants, or other living organisms threatened with extinction by anthropogenic (human-caused) or other natural changes in their environments. Requirements for declaring a species endangered are contained in the Endangered Species Act.

End-of-the-pipe equipment: Technologies such as scrubbers on smokestacks and catalytic converters on automobile tailpipes that reduce emissions of pollutants after they have formed.

Enforcement: EPA, state, or local legal actions to obtain compliance with environmental laws, rules, regulations, or agreements and/or obtain penalties or criminal sanctions for violations. Enforcement procedures may vary, depending on the requirements of different environmental laws and related implementing regulations.

Environment: All of the surrounding conditions and influences (physical and biological) affecting the development of living things; often refers to natural resources like air, land, and water.

Environmental audit: An independent assessment of the current status of a party's compliance with applicable environmental requirements or of a party's environmental compliance policies, practices and controls.

Environmental Impact Statement: A document required of federal agencies by the National Environmental Policy Act for major projects or legislative proposals significantly affecting the environment. A tool for decision making, it describes the positive and negative effects of an undertaking and cites alternative actions.

Environmental laws: Acts or statutes passed by the U.S. Congress and state legislatures to govern activities that pollute air, land, and water and to protect human health and the environment.

Environmental liability: The legal and financial responsibility of property owners to ensure that activities conducted on their properties, particularly in respect to regulated substances, are in compliance with applicable laws.

Environmental site assessment: The process of determining whether contamination is present on a parcel of real property.

Environmental toxicology: The science of harmful substances as they relate to human health and ecological impacts.

EPA Identification Number: The number assigned by EPA to each hazardous waste generator or transporter, and treatment, storage, or disposal facility.

Exposure: Contact with a chemical or contaminated media through inhalation, absorption, etc.

Exposure assessment: An investigation identifying the pathways by which toxicants may reach individuals, estimating how much of a chemical an individual is likely to be exposed to and estimating the number likely to be exposed.

Exposure pathway: The path from sources of pollutants via soil, water, or food to humans and other species or settings.

Extraction procedure (EP Toxic): A procedure that simulates leaching to determine toxicity; if a certain concentration of a toxic substance can be leached from a waste, that waste is considered hazardous, i.e., ''EP Toxic.''

Extremely hazardous substances: Any of the chemicals identified by EPA as toxic and listed under SARA Title III. The list is subject to periodic revision.

Facility emergency coordinator: Representative of a facility covered by environmental law (e.g., a chemical plant), who participates in the emergency reporting process with the local emergency planning committee (LEPC).

Feasibility study: (1). Analysis of the practicability of a proposal; e.g., a description and analysis of potential cleanup alternatives for a site. The feasibility study usually recommends selection of a cost-effective alternative. It generally starts as soon as a remedial investigation is under way; together, they are commonly referred to as the RI/FS. (2). A small-scale investigation of a problem to ascertain whether a proposed research approach is likely to provide useful data.

Fecal coliform bacteria: Bacteria found in the intestinal tracts of mammals. Their presence in water or sludge is an indicator of pollution and possible contamination by pathogens.

Financial assurance for closure: Documentation or proof that an owner or operator of a facility such as a landfill or other waste repository is capable of paying the projected costs of closing the facility and monitoring it afterward as provided in RCRA regulations.

Finding of No Significant Impact (FONSI): A document prepared by a federal agency stating why a proposed action would not have a significant impact on the environment and thus does not require preparation of an Environmental Impact Statement. An FNSI is based on the results of an environmental assessment.

Flammable: Any material that ignites easily and will burn rapidly.

Flash point: The lowest temperature at which evaporation of a substance produces sufficient vapor to form an ignitable mixture with air.

Flow rate: The rate, expressed in gallons (or liters) per hour, at which a fluid escapes from a hole or fissure in a tank. Such measurements are also made of liquid waste, effluent, and surface water movement.

Fluorocarbons (FCs): Any of a number of organic compounds analogous to hydrocarbons in which one or more hydrogen atoms are replaced by fluorine. Once used in the United States as a propellant for domestic aerosols, they are now found mainly in coolants and some industrial processes. FCs containing chlorine are called chlorofluorocarbons (CFCs). They are believed to be modifying the ozone layer in the stratosphere, thereby allowing more harmful solar radiation to reach the earth's surface.

Form R: Toxic chemical release inventory reporting form used to comply with Section 313 of SARA.

Friable: Capable of being crumbled, pulverized, or reduced to powder by hand pressure.

Friable asbestos: Any material containing more than 1% asbestos and that can be crumbled or reduced to powder by hand pressure. (May include previously nonfriable material that is broken or damaged by mechanical force.)

Fugitive emissions: Emissions not caught by a capture system.

General permit: A permit applicable to a class or category of dischargers.

Generator: A facility or mobile source that emits pollutants into the air or creates hazardous waste.

Global warming: An increase in the near surface temperature of the earth. Global warming has occurred in the distant past as the result of natural influences, but the term is most often used to refer to the warming predicted to occur as a result of increased emissions of greenhouse gases.

Good housekeeping: In this context, routine facility cleanup and maintenance to prevent environmental problems.

Grab sample: A single sample collected at a particular time and place that represents the composition of water, air, or soil only at that time and place.

Greenhouse effect: The warming of the Earth's atmosphere attributed to a buildup of carbon dioxide and/or other gases.

Greenhouse gas: A gas, such as carbon dioxide or methane, that contributes to potential climate change.

Ground water: The supply of fresh water found beneath the earth's surface, usually in aquifers, that supply wells and springs. Because groundwater is a major source of drinking water, there is growing concern over contamination from leaching agricultural or industrial pollutants or leaking underground storage tanks.

Habitat: The place where a population (e.g., human, animal, plant, microorganism) lives and its surroundings, both living and nonliving.

Hazard assessment: An investigation evaluating the effects of a stressor or determining a margin of safety for an organism by comparing the concentration that causes effects with an estimate of exposure to the organism.

Hazardous air pollutants: Air pollutants that are not covered by ambient air quality standards but which, as defined in the Clean Air Act, may present a threat of adverse human health effects or adverse environmental effects. Such pollutants include asbestos, beryllium, mercury, benzene, coke oven emissions, radionuclides, and vinyl chloride.

Hazardous and Solid Waste Amendments (HSWA): A law that revised the RCRA statutes to include notification and technical provisions for underground storage tanks and established land disposal bans for certain hazardous wastes.

Hazardous chemical: An EPA designation for any hazardous material requiring an MSDS under OSHA's Hazard Communication Standard.

Hazardous materials: Materials included in all the regulatory phrases that describe hazardous chemicals, such as hazardous substances (CERCLA), hazardous wastes (RCRA), hazardous materials (DOT), carcinogens (OSHA) and extremely hazardous substances (SARA).

Hazardous materials transportation act: A federal law that provides for the regulation of hazardous materials that are transported by air, water, rail, or highway.

Hazardous waste: Defined by the U.S. EPA under RCRA as a material that by definition is a *solid waste* that is either listed as hazardous or displays certain hazardous characteristics.

Hazardous waste landfill: An excavated or engineered site where hazardous waste is deposited and covered.

Hazard ratio: A ratio comparing an animal's daily dietary intake of a pesticide with its LD50 value. A ratio greater than 1.0 indicates that the animal is likely to consume a dose amount that would kill 50% of animals of the same species.

Health assessment: An evaluation of available data on existing or potential risks to human health posed by a Superfund site.

Heavy metals: Metallic elements with high atomic weights (e.g., mercury, chromium, cadmium, arsenic, and lead); can damage living things at low concentrations and tend to accumulate in the food chain.

Human health risk: The likelihood that a given exposure or series of exposures may have damaged or will damage the health of individuals.

Hydraulic gradients: The change in hydraulic head between two points divided by the distance between the points.

Hydrocarbons (HC): Chemical compounds that consist entirely of carbon and hydrogen.

Hydrogen sulfide (H_2.S): Gas emitted during organic decomposition. Also a by-product of oil refining and burning. Smells like rotten eggs and, in heavy concentration, can kill or cause illness.

Hydrology: The science dealing with the properties, distribution and circulation of water.

Identification Code or EPA I.D. Number: The unique code assigned to each generator, transporter and treatment, storage, or disposal facility by regulating agencies to facilitate identification and tracking of chemicals or hazardous waste.

Ignitable: Capable of burning or causing a fire.

Ignitable wastes: Wastes including any flammable or combustible liquid or solid.

Impermeable: Not easily penetrated. The property of a material or soil that does not allow, or allows only with great difficulty, the movement or passage of water.

Incidence rate: The number of new cases divided by the population at risk (i.e., one new incident or case in a population of 100,000 may be shown a $1 \times 10-5$.

In situ: In its original place; unmoved unexcavated; remaining at the site or in the subsurface.

In-situ technologies: Treatment technologies that can be applied without large-scale removal of the contaminated material from its existing location.

In vitro: Testing or action outside an organism (e.g., inside a test tube or culture dish).

In vivo: Testing or action inside an organism.

Incinerator: A furnace for burning waste under controlled conditions.

Industrial waste: Unwanted materials from an industrial operation; may be liquid, sludge, solid, or hazardous waste.

Influent: Water, wastewater, or other liquid flowing into a reservoir, basin, or treatment plant.

Injection wells: A well into which fluids are injected, primarily for disposal purposes.

Judicial or administrative orders: Orders that typically direct that certain actions be taken.

Lagoon: A shallow pond where sunlight, bacterial action, and oxygen work to purify wastewater; also used for storage of wastewater or spent nuclear fuel rods.

Land ban: Phasing out of land disposal of most untreated hazardous wastes, as mandated by the 1984 RCRA amendments.

Land disposal restrictions: Rules that require hazardous wastes to be treated before disposal on land to destroy or immobilize hazardous constituents that may migrate into soil and groundwater.

Landfilling: The final disposal of material that cannot be otherwise beneficially used, either because of the nature of the material or the local limitations to alternative approaches such as combustion.

Landfills: (1) Sanitary landfills are disposal sites for nonhazardous solid wastes, spread in layers, compacted to the smallest practical volume, and covered by material applied at the end of each operating day. (2) Secure chemical landfills are disposal sites for hazardous wastes, selected and designed to minimize the chance of release of hazardous substances into the environment.

LC 50/lethal concentration: Median-level concentration, a standard measure of toxicity. It tells how much of a substance is needed to kill half of a group of experimental organisms in a given time.

LD 50/lethal dose: The dose of a toxicant or microbe that will kill 50% of the test organisms within a designated period. The lower the LD 50, the more toxic the compound.

Leachate: Water that collects contaminants as it trickles through wastes, pesticides, or fertilizers. Leaching may occur in farming areas, feedlots, and landfills and may result in hazardous substances entering surface water, groundwater, or soil.

Leachate collection system: A system that gathers leachate and pumps it to the surface for treatment.

Leaching: The process by which soluble constituents are dissolved and filtered through the soil by a percolating fluid.

Lead (Pb): A heavy metal that is hazardous to health if breathed or swallowed. It is used in gasoline, paints, and plumbing.

Life cycle of a product: All stages of a product's development, from extraction of fuel for power to production, marketing, use, and disposal.

Lifetime exposure: Total amount of exposure to a substance that a human would receive in a lifetime (usually assumed to be 70 years).

Liner: A continuous layer of materials lining a surface impoundment, landfill, landfill cell, or container.

Local emergency planning committee (LEPC): A committee appointed by the state emergency response commission, as required by SARA Title III, to formulate a comprehensive emergency plan for its jurisdiction.

Lowest achievable emission rate: Under the Clean Air Act, the rate of emissions that reflects (1) the most stringent emissions limitation in the implementation plan of any state for such source unless the owner or operator demonstrates that such a limitation is not achievable, or (2) the most stringent emissions limitation achieved in practice, whichever is more stringent. A proposed new or modified source may not emit pollutants in excess of existing new source standards.

Lowest observable adverse effect level (LOAEL): The lowest level of a stressor that causes statistically and biologically significant differences in test samples as compared with other samples subjected to no stressor.

Major modification: This term is used to define modifications of major stationary sources of emissions in respect to ''Prevention of Significant Deterioration'' and ''New Source Review'' under the Clean Air Act.

Major stationary sources: Term used to determine the applicability of Prevention of Significant Deterioration and new source regulations.

Manifest: A one-page form used by haulers transporting waste, which lists EPA identification numbers, type and quantity of waste, the generator it originated from, the transporter that shipped it, and the storage or disposal facility to which it is being shipped. Includes copies for all participants in the shipping process.

Material Safety Data Sheet (MSDS): A compilation of information required under the OSHA Communication Standard on the identity of hazardous chemicals, health, physical hazards, exposure limits, and precautions.

Maximum Acceptable Toxic Concentration: For a given ecological effects test, the range (or geometric mean) between the No Observable Adverse Effect Level and the Lowest Observable Adverse Effects Level.

Maximum Achievable Control Technology (MACT): The emission standard for sources of air pollution requiring the maximum reduction of hazardous emissions, taking cost and feasibility into account.

Methanol: An alcohol that can be used as an alternative fuel or as a gasoline additive. It is less volatile than gasoline; when blended with gasoline it lowers carbon monoxide emissions but increases hydrocarbon emissions. Used as pure fuel, its emissions are less ozone-forming than those from gasoline. Poisonous to humans and animals if ingested.

Mineral: A naturally occurring solid with a defined crystal structure and a limited range of composition.

Modified source: The enlargement of a major stationary pollutant source is often referred to as modification, implying that more emissions will occur.

Monitoring: Periodic or continuous surveillance or testing to determine the level of compliance with statutory requirements and/or pollutant levels in various media or in humans, plants, and animals.

Monitoring well: A well used to obtain water quality samples or to measure groundwater levels.

National Ambient Air Quality Standards (NAAQS): Standards established by EPA that apply to outdoor air throughout the country.

National Emissions Standards for Hazardous Air Pollutants (NESHAPS): Emissions standards set by EPA for air pollutants not covered by NAAQS that may cause an increase in fatalities or in serious, irreversible, or incapacitating illness.

National Estuary Program: A program established under the Clean Water Act Amendments of 1987 to develop and implement conservation and management plans for protecting estuaries and restoring and maintaining their chemical, physical, and biological integrity, as well as controlling point and nonpoint pollution sources.

National Response Center: The federal operations center that receives notifications of all releases of oil and hazardous substances into the environment; open 24 hours a day; operated by the U.S. Coast Guard, which evaluates all reports and notifies the appropriate agencies.

Navigable waters: Waters of the United States that come under federal jurisdiction and are protected by certain provisions of the Clean Water Act.

Negligence: Conduct that falls below the standard established by law for the protection of others against unreasonable risk of harm.

Neutralization: Decreasing the acidity or alkalinity of a substance by adding alkaline or acidic materials, respectively.

New source: Any stationary source built or modified after publication of final or proposed regulations that establish a given standard of performance.

New Source Performance Standards (NSPS): Uniform national EPA air emission and water effluent standards, which limit the amount of pollution allowed from new sources or from modified existing sources.

New Source Review (NSR): A Clean Air Act requirement that State Implementation Plans must include a permit review that applies to the construction and operation of new and modified stationary sources in nonattainment areas to ensure attainment of National Ambient Air Quality Standards.

Nonattainment area: Area that does not meet one or more of the National Ambient Air Quality Standards for the criteria pollutants designated in the Clean Air Act.

Nonpoint Sources: Diffuse pollution sources (i.e., without a single point of origin or not introduced into a receiving stream from a specific outlet). The pollutants are generally carried off the land by storm water. Common nonpoint sources are agriculture, forestry, urban, mining, construction, dams, channels, land disposal, salt water intrusion, and city streets.

No Observable Adverse Effect Level: The dose at which there are no statistically or biologically significant increases in the frequency or severity of adverse effects observed in the study between the exposed population and control populations. Effects may be observed at this dose, but they are not considered adverse.

No Observable Effects Concentration: The highest concentration that causes no observed toxic effects.

No Observable Effect Level: The dose at which there are no statistically or biologically significant increases in the frequency or severity of the effects observed in the study between exposed and control populations.

Oil and gas waste: Gas and oil drilling muds, oil production brines, and other waste associated with exploration, development, and production of crude oil or natural gas.

Oil spill: An accidental or intentional discharge of oil. Oil spills to water bodies can be controlled by chemical dispersion, combustion, mechanical containment, and/or absorption.

Opacity: The amount of light obscured by particulate pollution in the air; clear window glass has zero opacity, a brick wall is 100% opaque. Opacity is an indicator of changes in performance of particulate control systems.

Oxidation: The chemical addition of oxygen to break down pollutants or organic waste, e.g., destruction of chemicals such as cyanides, phenols, and organic sulfur compounds in sewage by bacterial and chemical means.

Ozone depletion: Destruction of the stratospheric ozone layer that shields the earth from ultraviolet radiation harmful to life. This destruction of ozone is caused by the breakdown of certain chlorine-and/or bromine-containing compounds (chlorofluorocarbons or halons), which break down when they reach the stratosphere and then catalytically destroy ozone molecules.

Ozone layer: The protective layer in the atmosphere, about 15 miles above the ground, that absorbs some of the sun's ultraviolet rays, thereby reducing the amount of potentially harmful radiation that reaches the earth's surface.

Particulate loading: The mass of particulates per unit volume of air or water.

Particulates: (1) Fine liquid or solid particles such as dust, smoke, mist, fumes, or smog, found in air or emissions. (2) Very small solids suspended in water; these can vary in size, shape, density, and electrical charge and can be gathered together by coagulation and flocculation.

Pathogens: Microorganisms (e.g., bacteria, viruses, or parasites) that can cause disease in humans, animals, and plants.

Percolation: (1) The movement of water downward and radially through subsurface soil layers, usually continuing downward to groundwater. Can also involve upward movement of water. (2) Slow seepage of water though a filter.

Permeability: The rate at which liquids pass through soil or other materials in a specified direction.

Permissible dose: The dose of a chemical that may be received by an individual without the expectation of a significantly harmful result.

Permit: An authorization, license, or equivalent control document issued by EPA or an approved state agency to implement the requirements of an environmental regulation (e.g., a permit to operate a wastewater treatment plant or to operate a facility that may generate harmful emissions).

Personal protective equipment: Clothing and equipment worn by pesticide mixers, loaders, applicators, reentry workers, HAZMAT emergency responders, workers cleaning up Superfund sites, etc., which is worn to reduce their exposure to potentially hazardous chemicals and other pollutants.

pH: An expression of the intensity of the basic or acidic condition of a liquid; may range from 0 to 14, where 0 is the most acid and 7 is neutral. Natural waters usually have a pH between 6.5 and 8.5.

Phenols: Organic compounds that are by-products of petroleum refining, tanning, and textile, dye, and resin manufacturing. Low concentrations cause taste and odor problems in water; higher concentrations can kill aquatic life and humans.

Plug flow: Type of flow that occurs in tanks, basins, or reactors when a slug of water moves through without ever dispersing or mixing with the rest of the water flowing through.

Point source: A stationary location or fixed facility from which pollutants are discharged; any single identifiable source of pollution; (e.g., a pipe, ditch, ship, ore pit, factory, smokestack).

Pollutant: Generally, any substance introduced into the environment that adversely affects the usefulness of a resource or the health of humans, animals, or ecosystems.

Pollution prevention: The reduction in volume and/or toxicity of a waste prior to its being created or discharged.

Porosity: The degree to which an earth material is permeated with pores or cavities through which water or air can move.

Primary Drinking Water Regulation: Applies to public water systems and specifies a contaminant level, which, in the judgment of the EPA Administrator, will not adversely affect human health.

Primary waste treatment: First steps in wastewater treatment; screens and sedimentation tanks are used to remove most materials that float or will settle. Primary treatment removes about 30% of carbonaceous biochemical oxygen demand from domestic sewage.

Project XL: An EPA initiative to give states and the regulated community the flexibility to develop comprehensive strategies as alternatives to current regulatory requirements.

Protocol: A typical procedure for completing an environmental study, risk assessment, or other activity.

Public hearing: A formal meeting wherein EPA officials hear the public's views and concerns about an EPA action or proposal. EPA is required to consider such comments when evaluating its actions.

Publicly owned treatment works (POTW): A waste treatment works owned by a state, unit of local government, or Indian tribe, usually designed to treat domestic wastewaters and/or industrial effluents.

Public nuisance: A nuisance that involves interference with a general public right.

Public water supplies: Supplies defined by SDWA as those that serve at least 25 people for at least 60 days per year.

Public water system: A system that provides piped water for human consumption to at least 15 service connections or regularly serves 25 individuals.

Quality Assurance/Quality Control: A system of procedures, checks, audits, and corrective actions to ensure that all research design and performance, environmental monitoring and sampling, and other technical and reporting activities are of the highest achievable quality.

Radon: A colorless, naturally occurring, radioactive inert gas formed by radioactive decay of radium atoms in soil or rocks.

Raw sewage: Untreated wastewater and its contents.

Raw water: Intake water prior to any treatment or use.

Reactive: Characterizes hazardous waste that exhibits the following qualities:

- Reacts violently with water
- Is normally unstable and readily undergoes violent change without detonating
- Forms potentially explosive mixtures with water
- Is capable of explosion if subject to a strong initiating source

Recharge: The process by which water is added to a zone of saturation, usually by percolation from the soil surface, (e.g., the recharge of an aquifer).

Recycling: The on-site or off-site reutilization of a potential waste product.

Reference doses: Exposure limits for noncancer endpoints through the oral route of exposure.

Regulations: Rules written by federal or state agencies that define procedures for carrying out environmental laws.

Relative risk: The ratio between the risk in exposed subjects and the normal risk in a control population.

Release: Any spill, leaking, pumping, pouring, emitting, emptying, discharging, injecting, escaping, leaching, dumping, or disposing into the environment.

Remediation: Cleanup or other methods used to remove or contain a toxic spill or hazardous materials from a contaminated property.

Reportable quantity: For any CERCLA hazardous substance, the reportable quantity established and identified in the CERCLA list.

Residual risk: The extent of the health risk from air pollutants remaining after application of the Maximum Achievable Control Technology (MACT).

Ringlemann Chart: A series of shaded illustrations used to measure the opacity of air pollution emissions, ranging from light gray through black; used to set and enforce emissions standards.

Riparian rights: Entitlement of a landowner to certain uses of water on or bordering his or her property.

Risk assessment: Qualitative and quantitative evaluation of the risk posed to human health and/or the environment by the actual or potential presence and/or use of specific pollutants.

Risk characterization: The last phase of the risk assessment process that estimates the potential for adverse health or ecological effects to occur from exposure to a stressor and evaluates the uncertainty involved.

Runoff: Any liquid that drains over land from a facility; usually associated with storm water or discharges from nonpoint sources.

Safe Drinking Water Act (SDWA): A federal law intended to provide for the safety of drinking water supplies throughout the nation by establishing and enforcing national drinking water quality standards.

Sanitary landfill: A landfill where municipal garbage and other nonhazardous solid wastes are disposed of.

SARA: Superfund Amendments and Reauthorization Act; strengthens Superfund and includes provisions for emergency planning by communities to respond to releases of certain hazardous substances.

Secondary Drinking Water Regulations: Nonenforceable regulations applying to public water systems and specifying the maximum contamination levels that, in the judgment of EPA, are required to protect the public welfare. These regulations apply to any contaminants that may adversely affect the odor or appearance of such water and may consequently cause people served by the system to discontinue its use.

Secondary treatment: The second step in waste treatment systems in which bacteria consume the organic parts of the waste. It is accomplished by bringing together waste, bacteria, and oxygen in trickling filters or in the activated sludge process. This treatment removes floating and settleable solids and about 90% of the oxygen-demanding substances and suspended solids. Disinfection is the final state of secondary treatment.

Sediments: Soil, sand, and minerals washed from land into water, usually after rain. They pile up in reservoirs, rivers, and harbors, destroying fish and wildlife habitat and clouding the water so that sunlight cannot reach aquatic plants.

Sensitivity analysis: An analysis of the impact of varying parameters in a study.

Site reconnaissance: Visual and physical inspection of a property.

Site Safety and Health Plan: Required by OSHA in 29 CFR 1010.120 for hazardous waste sites; describes all potential hazards on-site and means for preventing employee exposure.

Small-quantity generator: A generator that generates more than 220 lb (100 kg) but less than 2,205 lb (1,000 kg) (or about 25 to less than 300 gal) of hazardous waste in a month and must comply with the hazardous waste management regulations.

Solvent extraction: A process that uses an organic solvent to separate hazardous organic contaminants from oily wastes, soils, sludges, and sediments, thereby reducing the volume of hazardous waste that must be treated.

Source: Equipment or activities that emit pollutants to the environment.

Source reduction: Reduction of the amount of materials entering a waste stream from a specific source by redesigning products or patterns of production or consumption (e.g., using returnable beverage containers). Synonymous with *waste reduction.*

Source separation: Segregation of various wastes at the point of generation (e.g., separation of paper, metal, and glass from other wastes to make recycling simpler and more efficient).

Statutory and regulatory requirements: Obligations directly imposed on facilities, apart from permits, orders, and schedules.

Superfund: The program operated under the legislative authority of CERCLA and SARA that funds and carries out EPA solid waste emergency and long-term removal and remedial activities. These activities include establishing the National Priorities List, investigating sites for inclusion on the list, determining their priority, and conducting and/or supervising cleanup and other remedial actions.

Surface impoundment: A facility formed primarily of earthen materials designed to hold liquid wastes (holding, settling, and aeration pits, ponds, lagoons).

Surface water: All water naturally open to the atmosphere (rivers, lakes, reservoirs, ponds, streams, impoundments, seas, estuaries, etc.).

Technology-based standards: Industry-specific effluent limitations applicable to direct and indirect sources, which are developed on a category-by-category basis using statutory factors.

Tertiary treatment: Advanced cleaning of wastewater that goes beyond the secondary or biological stage, removing nutrients such as phosphorus, nitrogen, and most BOD and suspended solids.

Threshold dose: A dose level below which no adverse effect is observable.

Total suspended solids (TSS): A measure of the suspended solids in wastewater, effluent, or water bodies, determined by tests for ''total suspended nonfilterable solids.''

Toxic pollutants: Materials that cause death, disease, or birth defects in organisms that ingest or absorb them. The quantities and exposures necessary to cause these effects can vary widely.

Toxics Release Inventory: Database of toxic releases in the United States compiled from SARA Title III Section 313 reports.

Toxic substance: A chemical or mixture that may present an unreasonable risk of injury to health or the environment.

Toxic waste: A waste than can produce injury if inhaled, swallowed, or absorbed through the skin.

Transporter: Any person engaged in the off-site transportation of hazardous waste by air, rail, highway, or water.

Treatment: Any method, technique, or process designed to remove solids and/or pollutants from solid waste, waste streams, effluents, or air emissions.

Turbidity: (1) Haziness in air caused by the presence of particles and pollutants. (2) A cloudy condition in water resulting from suspended silt or organic matter.

Underground injection wells: Steel and/or concrete–encased shafts into which hazardous wastes are deposited by force and under pressure.

Underground storage tank (UST): Any tank or combination of tanks (including underground pipes connected thereto) that is used to contain an accumulation of regulated substances and the volume of which is 10% or more beneath the surface of the ground.

Used oil: Spent motor oil from passenger cars and trucks collected at specified locations for recycling.

Vadose zone: The zone between land surface and the water table within which the moisture content is less than saturation (except in the capillary fringe) and pressure is less than atmospheric. Soil pore space typically also contains air or other gases. The capillary fringe is included in the vadose zone.

Vapor pressure: A measure of a substance's propensity to evaporate; vapor pressure is the force per unit area exerted by vapor in an equilibrium state with surroundings at a given pressure. It increases exponentially with an increase in temperature. A relative measure of chemical volatility, vapor pressure is used to calculate water partition coefficients and volatilization rate constants.

Viscosity: The molecular friction within a fluid that produces flow resistance.

Vitrification: The process of converting contaminated soil to a durable glasslike material in which the wastes are crystallized.

Waste: (1) Unwanted material left over from a manufacturing process. (2) Refuse from places of human or animal habitation.

Waste generation: The weight or volume of materials and products that enter the waste stream before recycling, composting, landfilling, or combustion takes place. Can also represent the amount of waste generated by a given source or category of sources.

Waste stream: The total flow of solid waste from homes, businesses, institutions, and manufacturing plants that is recycled, burned, or disposed of in landfills.

Wastewater: The spent or used water from a home, community, farm, or industry that contains dissolved or suspended matter.

Water pollution: The presence in water of enough harmful or objectionable material to damage the water's quality.

Water quality-based limitations: Effluent limitations applied to dischargers when mere technology-based limitations would cause violations of Water Quality Standards.

Water Quality Standards: State-adopted and EPA-approved ambient standards for water bodies. The standards prescribe the use of a water body and establish the water quality criteria that must be met to protect designated uses.

Watershed: The land area that drains into a stream; the watershed for a major river may encompass a number of smaller watersheds that ultimately combine at a common point.

Water table: The surface at which the water pressure in the pores of the porous geologic materials is exactly atmospheric.

Well: A bored, drilled, or driven shaft, or a dug hole, whose depth is greater than the largest surface dimension and whose purpose is to reach underground water supplies or oil, or to store or bury fluids below ground.

Wellhead protection area: A protected surface and subsurface zone surrounding a well or well field supplying a public water system to keep contaminants from reaching the well water.

Well point: A hollow vertical tube, rod, or pipe terminating in a perforated pointed shoe and fitted with a fine mesh screening.

Wetland: An area that is saturated by surface water or groundwater, with vegetation adapted for life under those soil conditions, such as swamps, bogs, fens, marshes, and estuaries.

Whole-Effluent-Toxicity Test: A test to determine the toxicity level of the total effluent from a single source, as opposed to a series of tests for individual contaminants.

APPENDIX C
EXAMPLE ENVIRONMENTAL FILING INDEX

	Retention File Code
ADMINISTRATION (COLOR CODED)	
Master File Index	—
Records Retention Schedule	—
Appropriation Requests	B
Budgets	A
Capital Plans	A
Commerce Department Surveys	B
Complaints Log	C
Compliance Orders	D
Consultants	A
Contracts	D
Correspondence—City	B
Correspondence—Corporate	A
Correspondence—County (including Health Department)	B
Correspondence—EPA	B
Correspondence—General	A
Correspondence—Legal	D
Correspondence—State Environmental Agency	B
Emergency Plans	C
Employee Records	A
Environmental Database	E
Environmental Plans	B
Environmental Tax Credits	D
Environmental Upset Reports	A
Expense Reports	A
Inspections—EPA	B
Inspections—State Environmental Agency	B
Invoices	A
Key Date Schedule	A
Labs	A
Notices of Violation	C
Performance Evaluations	C
Presentations	A
Property Assessments (Phase I)	E

Questionnaires	B
Research & Development	B
Spill Response Co-op Agreement	E
Studies and Evaluation	B
Surveys	A
Time Sheets	A
Trade Associations	A
Training Manuals	B
Training Records	B
Vendors and Suppliers	A

AIR (COLOR CODED)

Applications—NOC	C
Applications—PSD	C
Applications—Title V	C
Asbestos Plans	B
Asbestos Disposal Records	B
CEM(s) QA Plan	G
Compliance Certifications	C
Emissions—Fees	B
Emissions—Inventory	G
Dispersion Modeling	B
Met Station Data	D
Monitoring Data	D
Notifications to Agencies	D
Ozone-Depleting Chemicals	G
Permits—Historical	C
Permits—NOC	C
Permits—PSD	C
Permits—Title V	C
Regulations—Federal	G
Regulations—State	G
Reports	B
Source No. 1 Equipment Information	F
Source No. 2 Equipment Information	F
Source No. 3 Equipment Information	F
Source No. 4 Equipment Information	F
Source No. 5 Equipment Information	F
Source Test Manual	G
Source Test Reports	D
Test Certifications	C
Test Methods	B
Upset Log	C

HAZARDOUS WASTE (COLOR CODED)

Annual Report	G
Best Management Practices	G
Chemical Management	B
Form U—TSCA Chemical Inventory	B
Hazardous Waste Reduction Plan and Annual Report	B

PCB Records	B
Regulations—Federal	G
Regulations—State	G
Waste Analysis and Determinations	D
Waste Shipment Records	B

REPORTS AND PLANS (COLOR CODED)

Chemical Minimization Plan	A
Emergency Plans	G
Maintenance Plan	G
Oil Spill Prevention Plan	G
Pollution Prevention Plan	A
SARA Tier I/II Reports	B
SARA Tier III—Form R	B
SPCC Plan	G

SOLID WASTE (COLOR CODED)

Groundwater Monitoring	E
Groundwater Report	E
Groundwater Sampling and Analysis Plan	E
Land Application	B
Landfill Permit Fees	B
Landfill Permits	C
Landfill Quarterly Report	E
Leachate Analysis	D
Leachate Reports	B
Recycle	B
Sandblast Material Disposal	B
Solid Waste Management Plan	G
Solid Waste Minimization Plans and Studies	B
Treatment System Solids	D

WATER (COLOR CODED)

BMPs	G
Census Data (309)	C
Daily Data	B
Dredging	D
Effluent Analysis	D
Effluent Excursion Report	D
Effluent Guidelines	G
EPA 308 Requests	C
Groundwater Monitoring Data	E
Groundwater Studies	E
Industrial Pretreatment Permit with City	C
NPDES Permit	C
NPDES Permit Applications and Renewals	C
NPDES Reports	C
Outfall Inspections	B
Potable Water System (SDW)	C
Priority Pollutant Testing	C

Receiving Stream Information	B
Regulations—Federal	G
Regulations—State	G
Sanitary Sewer—Chlorine Residual Report	D
Secondary Treatment System	D
Spill Reports	B
Storm Water Annual Reports Testing/Fee	D
Storm Water Permit	C
Storm Water Pollution Control Plan	G
Water Rights	G
Water Use Certificate	E

Example
Retention Policy[a]

A—3 Years
B—5 Years
C—Settlement or Expiration +7 Years
D—10 Years
E—Permanent
F—Life of Equipment
G—Review Annually

[a] Actual retention policy should be reviewed with legal counsel for the facility or company.

APPENDIX D

GENERAL ENVIRONMENTAL AUDIT CHECKLIST

I. AIR POLLUTION

- Review compliance with air permit requirements, including new source performance standards if applicable.
- Review status of ambient air quality standards.
- Review applicable air monitoring requirements.
- Review emission of, or potential to emit, hazardous air pollutants.
- Review identification and management of asbestos.
- Review fugitive air emissions.
- Review compliance with opacity requirements if applicable.
- Determine whether there are any unpermitted emissions of air pollutants.
- Review inspection reports, notices of violations, and employee and public complaints.
- Review compliance with record-keeping and reporting requirements.

II. WATER POLLUTION

- Review compliance with NPDES permit requirements.
- Review storm water discharges.
- Determine whether there are any unpermitted water discharges.
- Discuss any water quality concerns.
- Review discharges of sanitary wastes.
- Review discharges to POTWs and compliance with any applicable pretreatment permit.

- Review dredge and fill permit issues.
- Discuss groundwater regulation and related permits if applicable.
- Review underground injection practices if applicable.
- Observe outfalls and wastewater treatment system.
- Review wastewater monitoring programs.
- Review spill handling and reporting procedures.
- Review compliance with record-keeping and reporting requirements.
- Review inspection reports, notices of violations, and employee and public complaints.

III. SOLID WASTE MANAGEMENT

- Review various solid waste streams and disposal or recycle locations.
- Review compliance with applicable solid waste disposal permits.
- Review disposal by other parties of solid waste on company property.
- Review inspection reports, notices of violation, and employee and public complaints.
- Review compliance with record-keeping and reporting requirements.

IV. HAZARDOUS WASTE MANAGEMENT

- Review various hazardous waste streams and disposal or recycling locations.
- Determine compliance with RCRA permit requirements if applicable.
- Determine compliance with hazardous waste generator requirements (e.g., compliance with storage, labeling, employee training requirements).
- Review management of waste oil, solvents, batteries, and waste paint.
- Discuss knowledge of past hazardous waste disposal practices.
- Review inspection reports, notices of violation, and employee and public complaints.
- Review manifests and other hazardous waste records.
- Review compliance with reporting requirements.

V. HAZARDOUS MATERIAL AND PETROLEUM STORAGE AND MANAGEMENT

- Inspect storage tanks for integrity and secondary containment.
- Determine whether an SPCC plan is required and, if so, ensure that the plan complies with applicable regulatory requirements.
- Review chemical transfer areas and standard operating procedures.
- Discuss spill response and reporting procedures.
- Review inspection reports, notices of violation, and employee and public complaints.
- Review record-keeping.
- Review notification and reporting procedures for spill events.

VI. UNDERGROUND STORAGE TANKS

- Identify all tanks currently on-site.
- Discuss past closure or removal of tanks.
- Review registration of tanks.
- Discuss methods of leak detection.
- Determine whether there is any evidence of leaking tanks.
- Discuss inspection reports, notices of violation, and employee and public complaints.

VII. TSCA AND PCB MANAGEMENT

- Identify whether PCB transformers or capacitors are present.
- If so, determine whether this equipment complies with applicable storage, labeling, inspection, and other requirements.
- Review disposal of PCB materials.
- Review inspection reports, notices of violation, and employee and public complaints.
- Review compliance with record-keeping and reporting requirements.

VIII. SAFE DRINKING WATER ACT

- Determine whether the facility is subject to the SDWA.
- If so, determine compliance with applicable regulatory requirements.

- Discuss inspection reports, notices of violation, and employee and public complaints.
- Review compliance with testing, record-keeping, and reporting requirements.

IX. EMERGENCY PLANNING AND COMMUNITY RIGHT-TO-KNOW LAW

- Determine whether the facility is subject to the Community Right-to-Know Law.
- If so, review MSDS reporting requirements.
- Review toxic chemical release reporting requirements.
- Review inspection reports, notices of violations, and complaints.
- Review compliance with record-keeping and reporting requirements.
- Review emergency response plan.
- Review emergency release notification procedures.

X. ENVIRONMENTAL MANAGEMENT PROGRAMS

- Determine whether the facility has a written environmental policy.
- Review the facility's Environmental Management System.
- Review environmental filing and record-keeping procedures.
- Examine environmental training materials and training records.
- Determine how environmental planning is integrated into the facility's capital planning process.
- Review standard operating procedures.
- Examine the facility's pollution prevention program.
- Discuss the facility's public involvement/community outreach efforts.

APPENDIX E

WASHINGTON STATE ENVIRONMENTAL POLICY ACT ENVIRONMENTAL CHECKLIST

PURPOSE OF CHECKLIST

The State Environmental Policy Act (SEPA), chapter 43.21C RCW, requires all governmental agencies to consider the environmental impacts of a proposal before making decisions. An environmental impact statement (EIS) must be prepared for all proposals with probable significant adverse impacts on the quality of the environment. The purpose of this checklist is to provide information to help you and the agency identify impacts from your proposal (and to reduce or avoid impacts from the proposal, if it can be done) and to help the agency decide whether an EIS is required.

INSTRUCTIONS FOR APPLICANTS

This environmental checklist asks you to describe some basic information about your proposal. Governmental agencies use this checklist to determine whether the environmental impacts of your proposal are significant, requiring preparation of an EIS. Answer the questions briefly, with the most precise information known, or give the best description you can.

You must answer each question accurately and carefully, to the best of your knowledge. In most cases, you should be able to answer the questions from your own observations or project plans without the need to hire experts. If you really do not know the answer, or if a question does not apply to your proposal, write "Do not know" or "Does not apply." Complete answers to the questions now may avoid unnecessary delays later.

Some questions ask about governmental regulations, such as zoning, shoreline, and landmark designations. Answer these questions if you can. If you have problems, the governmental agencies can assist you.

The checklist questions apply to all parts of your proposal, even it you plan to do them over a period of time or on different parcels of land. Attach any additional information that will help describe your proposal or its environmental effects. The agency to which you submit this checklist may ask you to explain your answers or provide additional information reasonably related to determining whether there may be significant adverse impact.

USE OF CHECKLIST FOR NONPROJECT PROPOSALS

Complete this checklist for nonproject proposals, even though questions may be answered "does not apply."

For nonproject actions, the references in the checklist to the words "project," "applicant," and "property or site" should be read as "proposal," "proposer," and "affected geographic area," respectively.

A. Background

1. Name of proposed project, if applicable:
2. Name of applicant:
3. Address and phone number of applicant and contact person:
4. Date checklist prepared:
5. Agency requesting checklist:
6. Proposed timing or schedule (including phasing, if applicable):
7. Do you have any plans for future additions, expansion, or further activity related to or connected with this proposal? If yes, explain.
8. List any environmental information you know about that has been prepared, or will be prepared, directly related to this proposal.
9. Do you know whether applications are pending for government approvals of other proposals directly affecting the property covered by your proposal? If yes, explain.
10. List any government approvals or permits that will be needed for your proposal, if known.
11. Give a brief, complete description of your proposal, including the proposed uses and the size of the project and site. There are several questions later in this checklist that ask you to describe certain aspects of your proposal. You do not need to repeat those answers on this page. (Lead agencies may modify this form to include additional specific information on project description.)

12. Location of the proposal. Give sufficient information for a person to understand the precise location of your proposed project, including a street address, if any, and section, township, and range, if known. If the proposal would occur over a range of area, provide the range or boundaries of the site(s). Provide a legal description, site plan, vicinity map, and topographic map, if reasonably available. Although you should submit any plans required by the agency, you are not required to duplicate maps or detailed plans submitted with any permit applications related to this checklist.

To Be Completed by Applicant	Evaluation for Agency Use Only
B. Environmental Elements **1. Earth** a. General description of the site (circle one): Flat, rolling, hilly, steep slopes, mountainous, other __________. b. What is the steepest slope on the site (approximate percent slope)? c. What general types of soils are found on the site (for example, clay, sand, gravel, peat, muck)? If you know the classification of agricultural soils, specify them and note any prime farmland. d. Are there surface indications or history of unstable soils in the immediate vicinity? If so, describe. e. Describe the purpose, type, and approximate quantities of any filling or grading proposed. Indicate source of fill. f. Could erosion occur as a result of clearing, construction, or use? If so, generally describe. g. About what percentage of the site will be covered with impervious surfaces after project construction (for example, asphalt or buildings)? h. Proposed measures to reduce or control erosion, or other impacts to the earth, if any: **2. Air** a. What types of emissions to the air would result from the proposal (i.e., dust, automobiles, odors, industrial wood smoke) during construction and when the project is completed? If any, generally describe and give approximate quantities if known.	

To Be Completed by Applicant	Evaluation for Agency Use Only

b. Are there any off-site sources of emissions or odor that may affect your proposal? If so, generally describe.

c. Proposed measures to reduce or control emissions or other impacts to air, if any:

3. Water

a. Surface

(1) Is there any surface water body on or in the immediate vicinity of the site (including year-round and seasonal streams, salt water, lakes, ponds, wetlands)? If yes, describe type and provide name. If appropriate, state what stream or river it flows into.

(2) Will the project require any work over, in, or adjacent to (within 200 feet) the described waters? If yes, please describe and attach available plans.

(3) Estimate the amount of fill and dredge material that would be placed in or removed from surface water or wetlands, and indicate the area of the site that would be affected. Indicate the source of fill material.

(4) Will the proposal require surface water withdrawals or diversions? Give general description, purpose, and approximate quantities if known.

(5) Does the proposal lie within a 100-year floodplain? If so, note location on the site plan.

(6) Does the proposal involve any discharges of waste materials to surface waters? If so, describe the type of waste and anticipated volume of discharge.

b. Ground

(1) Will groundwater be withdrawn, or will water be discharged to groundwater? Give general description, purpose, and approximate quantities if known.

(2) Describe wastewater that will be discharged into the ground from septic tanks or other sources, if any (for example: Domestic sewage: Industrial, containing the following chemicals . . . ; agricultural; etc.). Describe the general size of the system, the number of such systems, the number of houses to be served (if applicable), or the number of animals or humans the system(s) are expected to serve.

To Be Completed by Applicant	Evaluation for Agency Use Only

c. Water runoff (including storm water)

(1) Describe the source of runoff (including storm water) and method of collection or disposal, if any (include quantities, if known). Where will this water flow? Will this water flow into other waters? If so, describe.

(2) Could waste materials enter ground or surface waters? If so, generally describe.

d. Proposed measures to reduce or control surface, ground, and runoff water impacts, if any:

4. Plants

a. Check or circle types of vegetation found on the site:
__ deciduous tree: alder, maple, aspen, other
__ evergreen tree: fir, cedar, pine, other
__ shrubs
__ grass
__ pasture
__ crop or grain
__ wet soil plants: cattail, buttercup, bullrush, skunk cabbage, other
__ water plants: water lily, eelgrass, milfoil, other
__ other types of vegetation

b. What kind and amount of vegetation will be removed or altered?

c. List threatened or endangered species known to be on or near the site.

d. Proposed landscaping, use of native plants, or other measures to preserve or enhance vegetation on the site, if any:

5. Animals

a. Circle the names of any birds and animals that have been observed on or near the site or are known to be on or near the site:

birds: hawk, heron, eagle, songbirds, other: ________
mammals: deer, bear, elk, beaver, other: ________
fish: bass, salmon, trout, herring, shellfish, other: ________

b. List any threatened or endangered species known to be on or near the site.

To Be Completed by Applicant	**Evaluation for Agency Use Only**

c. Is the site part of a migration route? If so, explain.

d. Proposed measures to preserve or enhance wildlife, if any:

6. Energy and Natural Resources

a. What kinds of energy (electric, natural gas, oil, wood stove, solar) will be used to met the completed project's energy needs? Describe whether it will be used for heating, manufacturing, etc.

b. Would your project affect the potential use of solar energy by adjacent properties? If so, generally describe.

c. What kinds of energy conservation features are included in the plans of this proposal? List other proposed measures to reduce or control energy impact, if any:

7. Environmental Health

a. Are there any environmental health hazards, including exposure to toxic chemicals, risk of fire and explosion, spill, or hazardous waste, that could occur as a result of this proposal? If so, describe.

(1) Describe special emergency services that might be required.

(2) Proposed measures to reduce or control environmental health hazards, if any:

b. Noise

(1) What types of noise exist in the area which may affect your project (for example: traffic, equipment, operation, other)?

(2) What types and levels of noise would be created by or associated with the project on a short-term or a long-term basis (for example: traffic, construction, operation, other)? Indicate what hours noise would come from the site.

(3) Proposed measures to reduce or control noise impacts, if any:

8. Land and Shoreline Use

a. What is the current use of the site and adjacent properties?

b. Has the site been used for agriculture? If so, describe.

c. Describe any structures on the site.

To Be Completed by Applicant	**Evaluation for Agency Use Only**

d. Will any structures be demolished? If so, what?
e. What is the current zoning classification of the site?
f. What is the current comprehensive plan designation of the site?
g. If applicable, what is the current shoreline master program designation of the site?
h. Has any part of the site been classified as an "environmentally sensitive" area? If so, specify.
i. Approximately how many people would reside or work in the completed project?
j. Approximately how many people would the completed project displace?
k. Proposed measures to avoid or reduce displacement impacts, if any:
l. Proposed measures to ensure that the proposal is compatible with existing and projected land uses and plans, if any:

9. Housing

a. Approximately how many units would be provided, if any? Indicate whether high-, middle-, or low-income housing.
c. Proposed measures to reduce or control housing impacts, if any:

10. Aesthetics

a. What is the tallest height of any proposed structure(s), not including antennas; what is the principal exterior building material(s) proposed?
b. What views in the immediate vicinity would be altered or obstructed?
c. Proposed measures to reduce or control aesthetic impacts, if any:

11. Light and Glare

a. What type of light or glare will the proposal produce? What time of day would it mainly occur?
b. Could light or glare from the finished project be a safety hazard or interfere with views?
c. What existing off-site sources of light or glare may affect your proposal?

To Be Completed by Applicant	Evaluation for Agency Use Only

d. Proposed measures to reduce or control light and glare impacts, if any:

12. Recreation

a. What designated and informal recreational opportunities are in the immediate vicinity?

b. Would the proposed project displace any existing recreational uses? If so, describe.

c. Proposed measures to reduce or control impacts on recreation, including recreation opportunities to be provided by the project or applicant, if any:

13. Historical and Cultural Preservation

a. Are there any places or objects listed on, or proposed for, national, state, or local preservation registers known to be on or next to the site? If so, generally describe.

b. Generally describe any landmarks or evidence of historical, archaeological, scientific, or cultural importance known to be on our next to the site.

c. Proposed measures to reduce or control impacts, if any:

14. Transportation

a. Identify public streets and highways servicing the site, and describe proposed access to the existing street system. Show on site plans, if any.

b. Is site currently served by public transit? If not, what is the appropriate distance to the nearest transit stop?

c. How many parking spaces would the completed project have? How many would the project eliminate?

d. Will the proposal require any new roads or streets, or improvements to existing roads or streets, not including driveways? If so, generally describe (indicate whether public or private).

e. Will the project use (or occur in the immediate vicinity of) water, rail, or air transportation? If so, generally describe.

f. How many vehicular trips per day would be generated by the completed project? If known, indicate when peak volumes would occur.

g. Proposed measures to reduce or control transportation impacts, if any:

To Be Completed by Applicant	Evaluation for Agency Use Only

15. Public Services

a. Would the project result in an increased need for public services (for example, fire protection, police protection, health care, schools, other)? If so, generally describe.

b. Proposed measures to reduce or control direct impacts on public services, if any.

16. Utilities

a. Circle utilities currently available at the site: electricity, natural gas, water, refuse service, telephone, sanitary sewer, septic system, other.

b. Describe the utilities that are proposed for the project, the utilities providing the services, and the general construction activities on the site or in the immediate vicinity that might be needed.

C. Signature

The above answers are true and complete to the best of my knowledge. I understand that the lead agency is relying on them to make its decision.

Signature: ________________

Date Submitted: ____________

APPENDIX F

NPDES PERMIT GENERAL CONDITIONS

G1. DISCHARGE VIOLATIONS

All discharges and activities authorized by this permit shall be consistent with the terms and conditions of this permit. The discharge of any pollutant more frequently than, or at a concentration in excess of, that authorized by this permit shall constitute a violation of the terms and conditions of this permit.

G2. PROPER OPERATION AND MAINTENANCE

The permittee shall at all times properly operate and maintain all facilities and systems of collection, treatment, and control (and related appurtenances) which are installed or used by the permittee for pollution control.

G3. REDUCED PRODUCTION FOR COMPLIANCE

The permittee, in order to maintain compliance with its permit, shall control production and/or all discharges upon reduction, loss, failure, or bypass of the treatment facility until the facility is restored or an alternative method of treatment is provided. This requirement applies in the situation where, among other things, the primary source of power of the treatment facility is reduced, lost, or fails.

G4. NONCOMPLIANCE NOTIFICATION

If, for any reason, the pemittee does not comply with, or will be unable to comply with, any of the discharge limitations or other conditions specified in the permit, the permittee shall, at a minimum, provide the agency with the following information:

a. A description of the nature and cause of noncompliance, including the quantity and quality of any unauthorized waste discharges;
b. The period of noncompliance, including exact dates and times and/or the anticipated time when the permittee will return to compliance; and
c. The steps taken, or to be taken, to reduce, eliminate, and prevent recurrence of the noncompliance.

In addition, the permittee shall take immediate action to stop, contain, and clean up any unauthorized discharges and take all reasonable steps to minimize any adverse impacts to waters of the state and correct the problem. The permittee shall notify the department by telephone so that an investigation can be made to evaluate any resulting impacts and the corrective actions taken to determine if additional action should be taken.

In the case of any discharge subject to any applicable toxic pollutant effluent standard under Section 307(a) of the Clean Water Act, or which could constitute a threat to human health, welfare, or the environment, 40 CFR Part 122 requires that the information specified in items G4.a., G4.b., and G4.c, shall be provided not later than 24 hours from the time the permittee becomes aware of the circumstances. If this information is provided orally, a written submission covering these points shall be provided within five days of the time the permittee becomes aware of the circumstances, unless the agency waives or extends this requirement on a case-by-case basis.

Compliance with these requirements does not relieve the permittee from responsibility to maintain continuous compliance with the conditions of this permit or the resulting liability for failure to comply.

G5. BYPASS PROHIBITED

The intentional bypass of wastes from all or any portion of a treatment works is prohibited unless the following four conditions are met:

a. Bypass is: (1) unavoidable to prevent loss of life, personal injury, or severe property damage; or (2) necessary to perform construction or maintenance-related activities essential to meet the requirements of the Clean Water Act and authorized by administrative order.

b. There are no feasible alternatives to bypass, such as the use of auxiliary treatment facilities, retention of untreated wastes, maintenance during normal periods of equipment downtime, or temporary reduction or termination of production.

c. The permittee submits notice of an unanticipated bypass to the agency in accordance with Condition G4. Where the permittee knows or should have known in advance of the need for a bypass, this prior notification shall be submitted for approval to the agency, if possible, at least 30 days before the date of bypass (or longer if specified in the special conditions).

d. The bypass is allowed under conditions determined to be necessary by the agency to minimize any adverse effects. The public shall be notified and given an opportunity to comment on bypass incidents of significant duration, to the extent feasible.

''Severe property damage'' means substantial physical damage to property, damage to the treatment facilities which would cause them to become inoperable, or substantial and permanent loss of natural resources which can reasonably be expected to occur in the absence of a bypass. Severe property damage does not mean economic loss caused by delays in production.

After consideration of the factors above and the adverse effects of the proposed bypass, the agency will approve or deny the request. Approval of a request to bypass will be by administrative order.

G6. RIGHT OF ENTRY

The permittee shall allow an authorized representative of the agency, upon the presentation of credentials and such other documents as may be required by law:

a. To enter upon the premises where a discharge is located or where any records must be kept under the terms and conditions of this permit;

b. To have access to and copy at reasonable times any records that must be kept under the terms of the permit;

c. To inspect at reasonable times any monitoring equipment or method of monitoring required in the permit;

d. To inspect at reasonable times any collection, treatment, pollution management, or discharge facilities; and

e. To sample at reasonable times any discharge of pollutants.

G7. PERMIT MODIFICATIONS

The permittee shall submit a new application or supplement to the previous application where facility expansions, production increases, or process modifications will (1) result in new or substantially increased discharges of pollutants or a change in the nature of the discharge of pollutants, or (2) violates the terms and conditions of this permit.

G8. PERMIT MODIFIED OR REVOKED

After notice and opportunity for public hearing, this permit may be modified, terminated, or revoked during its term for cause as follows:

a. Violation of any terms or conditions of the permit;
b. Failure of the permittee to disclose fully all relevant facts or misrepresentations of any relevant facts by the permittee during the permit issuance process;
c. A change in any conditions that requires either a temporary or a permanent reduction or elimination of any discharge controlled by the permit;
d. Information indicating that the permitted discharge poses a threat to human health or welfare;
e. A change in ownership or control of the source; or
f. Other causes listed in 40 CFR Parts 122.62 and 122.63.

Permit modification, revocation and reissuance, or termination may be initiated by the agency or requested by any interested person.

G9. REPORTING A CAUSE FOR MODIFICATION

A permittee who knows or has reason to believe that any activity has occurred or will occur which would constitute cause for modification or revocation and reissuance under condition G8 or 40 CFR Part 122.62 must report such plans, or such information, to the agency so that a decision can be made on whether action to modify or revoke and reissue a permit will be required. The agency may then require submission of a new application. Submission of such application does not relieve the discharger of the duty to comply with the existing permit until it is modified or reissued.

G10. TOXIC POLLUTANTS

If any applicable toxic effluent standard or prohibition (including any schedule of compliance specified in such effluent standard or prohibition) is established under Section 307(a) of the Clean Water Act for a toxic pollutant and that standard or prohibition is more stringent than any limitation upon such pollutant in the permit, the agency shall institute proceedings to modify or revoke and reissue the permit to conform to the toxic effluent standard of prohibition.

G11. PLAN REVIEW REQUIRED

Prior to constructing or modifying any wastewater control facilities, detailed plans shall be submitted to the agency for approval. Facilities shall be constructed and operated in accordance with the approved plan.

G12. OTHER REQUIREMENTS OF 40 CFR

All other requirements of 40 CFR Parts 122.41 and 122.42 are incorporated in this permit by reference.

G13. COMPLIANCE WITH OTHER LAWS AND STATUTES

Nothing in the permit shall be construed as excusing the permittee from compliance with any applicable federal, state, or local statutes, ordinances, or regulations.

G14. ADDITIONAL MONITORING

The agency may establish specified monitoring requirements in addition to those contained in this permit by administrative order or permit modification.

G15. REVOCATION FOR NONPAYMENT OF FEES

The agency may revoke this permit if the permit fees are not paid.

G16. REMOVED SUBSTANCES

Collected screenings, grit, solids, sludges, filter backwash, or other pollutants removed in the course of treatment of wastewaters or control of wastewaters shall not be resuspended or reintroduced to the final effluent stream for discharge to state waters.

G17. DUTY TO REAPPLY

The permittee must reapply for a permit at least 180 days before the expiration of this permit.

APPENDIX G

EFFLUENT GUIDELINES EXAMPLE: ORGANIC PESTICIDE CHEMICALS, MANUFACTURING SUBCATEGORY—40 CFR §455.11 ET SEQ.

§455.22 Effluent limitations guidelines representing the degree of effluent reduction attainable by the application of the best practicable control technology currently available.

Except as provided in §§125.30 through 125.32, any existing point source subject to this subpart shall achieve the following effluent limitations representing the degree of effluent reduction attainable by the application of the best practicable control technology currently available (BPT). The following limitations establish the quantity or quality of pollutants or pollutant properties controlled by this paragraph which may be discharged from the manufacture of organic active ingredient:

	Effluent Limitations	
Effluent Limitations	Maximum for Any 1 Day	Average of Daily Values for 30 Consecutive Days Shall not Exceed
COD	13.000	9.0000
BOD5	7.400	1.6000
TSS	6.100	1.8000
Organic pesticide chemicals	.010	.0018
pH	[1]	[1]

[1] Within the range of 6.0 to 9.0

Note: For COD, BOD5, and TSS, metric units: kilogram/1,000 kg of total organic active ingredients. English units: pound/1,000 lb of total organic active ingredients. For organic pesticide chemicals, metric units: kilogram/1,000 kg of organic pesticide chemicals. English units: pound/1,000 lb of organic pesticide chemicals.

[43 FR 44845, Sept. 29, 1978, as amended at 60 FR 33971, June 29, 1995]

§455.23 Effluent limitations guidelines representing the degree of effluent reduction attainable by the application of the best conventional pollutant control technology (BCT).

Except as provided in 40 CFR 125.30 through 125.32, any existing point source subject to this subpart must achieve the effluent limitations representing the degree of effluent reduction attainable by the application of the best conventional pollutant control technology. The limitations for BOD, TSS, and pH are the same as those specified in 40 CFR 455.22.

BCT effluent limitations

Pollutant or Pollutant Property	Maximum for Any One Day**	Average of Daily Values Shall Not Exceed**
BOD5	7.400	1.6000
TSS	6.100	1.8000
pH	*	*

* Within the range of 6.0 to 9.0.
** Metric units: kilogram pollutant/1,000 kg of total organic active ingredients. English units: pound pollutant/1,000 lb of total organic active ingredients.

[58 FR 50689, Sept. 28, 1993]

§455.24 Effluent limitations guidelines representing the degree of effluent reduction attainable by the application of the best available control technology economically achievable (BAT).

Except as provided in 40 CFR 125.30 through 125.32, any existing point source subject to this subpart must achieve the effluent limitations representing the degree of effluent reduction attainable by the application of the best available technology as specified in 40 CFR 455.20(d). For the priority pollutants, such sources must achieve discharges not exceeding the quantity (mass) determined by multiplying the process wastewater flow subject to this subpart as defined in 40 CFR 455.21 (d) times the concentrations listed in Table 4 or Table 5 of this part, as appropriate, of this subpart.

[58 FR 50690, Sept. 28, 1993]

§455.25 New source performance standards (NSPS).

(a) Any new source subject to this subpart which discharges process wastewater pollutants must achieve the new source performance standards specified in 40 CFR 455.20(d), and subject to 455.20(a), must meet the following standards for BOD5, TSS, COD, and pH:

(b) For the priority pollutants, such sources must achieve discharges not exceeding the quantity (mass) determined by multiplying the process wastewater flow subject to this subpart as defined in 40 CFR 455.21(d) times the concentrations listed in Table 4 or Table 5 of this part, as appropriate, of this subpart.

New Source Performance Standards

Pollutant or Pollutant Property	Maximum for Any One Day**	Average of Daily Values Shall Not Exceed**
COD	9.360	6.480
BOD5	5.328	1.1520
TSS	4.392	1.2960
pH	*	*

*Within the range of 6.0 to 9.0.

**Metric units: kilogram pollutant/1,000 kg of total organic active ingredients. English units: pound pollutant/1,000 lb of total organic active ingredients.

[58 FR 50690, Sept. 28, 1993]

§455.26 Pretreatment standards for existing sources (PSES).

Except as provided in 40 CFR 403.7, any existing source subject to this subpart which introduces pollutants into a publicly owned treatment works must comply with 40 CFR part 403 and achieve the pretreatment standards for existing sources (PSES) as specified in 40 CFR 455.20(d). For the priority pollutants, such sources must achieve discharges not exceeding the quantity (mass) determined by multiplying the process wastewater flow subject to this subpart as defined in 40 CFR 4455.21(d) times the concentrations listed in Table 6 of this part. If mass limitations have not been developed as required, the source shall achieve discharges not exceeding the concentration limitations listed in Table 6 of this part.

[58 FR 50690, Sept. 28, 1993]

§455.27 Pretreatment standards for new sources (PSNS).

Except as provided in 40 CFR 403.7, any new source subject to this subpart which introduces pollutants into a publicly owned treatment works must comply with 40 CFR part 403 and must adhieve the pretreatment standards for new sources (PSNS) as specified in 40 CFR 455.20(d). For the priority pollutants, the source must achieve discharges not exceeding the quantity (mass) determined by multiplying the process wastewater flow subject to this subpart as defined in 40 CFR 455.21(d) times the concentrations listed in Table 6 of this part. If mass limitations have not been developed as required, the source shall achieve discharges not exceeding the concentration limitations listed in Table 6 of this part.

[58 FR 50690, Sept. 28, 1993]

APPENDIX H

NOTIFICATION OF DEMOLITION AND RENOVATION—40 CFR §61.45

Notification of Demolition and Renovation

40 CFR §61.45

Operator Project No.	Postmark	Date Received	Notification No.
I. Type of Notification (O=Original R=Revised C=Cancelled):			
II. Facility Information (Identify owner, removal contractor, and other operator)			
Owner Name:			
Address:			
City:	State:	Zip:	
Contact:		Tel:	
Removal Contractor:			
Address:			
City:	State:	Zip:	
Contact:		Tel:	
Other Operator:			
Address:			
City:	State:	Zip:	
Contact:		Tel:	
III. Type of Operation (D =Demo O=Ordered Demo R=Renovation E=Emer. Renovation):			
IV. Is Asbestos Present? (Yes/No)			
V. Facility Description (Include building name, number and floor or room number)			
Bldg. Name:			
Address:			
City:	State:	County:	
Site Location:			
Building Size:	No. of Floors:	Age in Years:	
Present Use:	Prior Use:		

VI. Procedure, Including Analytical Method, If Appropriate, Used to Detect the Presence of Asbestos Material:

VII. Approximate Amount of Asbestos, Including: 1. Regulated ACM to be removed 2. Category I ACM Not Removed 3. Category II ACM Not Removed	RACM To Be Removed	Nonfriable Asbestos Material Not To Be Removed		Indicate Unit of Measurement Below	
		Cat I	Cat II	Unit	
Pipes				LnFt:	Ln m:
Surface Area				SqFt:	Sq m:
Vol. RACM Off Facility Component				CuFt:	Cu m:

VIII. Scheduled Dates Asbestos Removal (mm/dd/yy) Start: Complete:
IX. Scheduled Dates Demo/Renovation (mm/dd/yy) Start: Complete:
X. Description of Planned Demolition or Renovation Work, and Method(s) to be Used:
XI. Description Work Practices and Engineering Controls to be Used to Prevent Emissions of Asbestos at the Demolition and Renovation Site:

XII. Waste Transporter #1		
Name:		
Address:		
City:	State:	Zip:
Contact Person:		Tel:
Waste Transporter #2		
Name:		
Address:		
City:	State:	Zip:
Contact Person:		Tel:
XIII. Waste Disposal Site		

Name:		
Location:		
City:	State:	Zip:
Telephone:		

XIV. If Demolition Ordered by a Government Agency, Please Identify the Agency Below:

Name:	Title:
Authority:	
Date of Order (mm/dd/yy):	Date Ordered to Begin (mm/dd/yy):

XV. For Emergency Renovations

Date and Hour of Emergency (mm/dd/yy):

Description of the Sudden, Unexpected Event:

Explanation of how the event caused unsafe conditions or would cause equipment damage or an unreasonable financial burden:

XVI. Description of Procedures to be Followed in the Event That Unexpected Asbestos is Found or Previously Nonfriable Asbestos Material Becomes Crumbled, Pulverized, or Reduced to Powder.

XVII. I Certify That an Individual Trained in the Provisions of this Regulation (40 CFR Part 61, Subpart M) Will be On-Site During the Demolition or Renovation and Evidence that the Required Training Has Been Accomplished by this Person will be Available for Inspection During Normal Business Hours. (Required one year after promulgation)

(Signature of Owner/Operator) (Date)

XVIII. I Certify That the Above Information is Correct.

(Signature of Owner/Operator) (Date)

APPENDIX I

EPA FORM R

(IMPORTANT: Type or print; read instructions before completing form)

Form Approved OMB Number: 2070-0093
Approval Expires: 04/2000

Page 1 of 5

EPA United States Environmental Protection Agency

FORM R

Section 313 of the Emergency Planning and Community Right-to-Know Act of 1986, also known as Title III of the Superfund Amendments and Reauthorization Act

TOXIC CHEMICAL RELEASE INVENTORY REPORTING FORM

WHERE TO SEND COMPLETED FORMS: 1. EPCRA Reporting Center, P.O. Box 3348, Merrifield, VA 22116-3348, ATTN: TOXIC CHEMICAL RELEASE INVENTORY 2. APPROPRIATE STATE OFFICE (See instructions in Appendix F)

Enter "X" here if this is a revision

For EPA use only

IMPORTANT: See instructions to determine when "Not Applicable (NA)" boxes should be checked.

PART I. FACILITY IDENTIFICATION INFORMATION

SECTION 1. REPORTING YEAR 19 ___

SECTION 2. TRADE SECRET INFORMATION

2.1	**Are you claiming the toxic chemical identified on page 2 trade secret?** ☐ **Yes (Answer question 2.2; Attach substantiation forms)** ☐ **No** Do not answer 2.2; go to Section 3	2.2	**Is this copy** ☐ **Sanitized** ☐ **Unsanitized** (Answer only if "YES" in 2.1)

SECTION 3. CERTIFICATION (Important: Read and sign after completing all form sections.)

I hereby certify that I have reviewed the attached documents and that, to the best of my knowledge and belief, the submitted information is true and complete and that the amounts and values in this report are accurate based on reasonable estimates using data available to the preparers of this report.

Name and official title of owner/operator or senior management official:	Signature:	Date signed:

SECTION 4. FACILITY IDENTIFICATION

TRI Facility ID Number

4.1	Facility or Establishment Name	Facility or Establishment Name or Mailing Address (if different from street address)
	Street	Mailing Address
	City/County/State/Zip Code	City/County/State/Zip Code

4.2	This report contains information for: (Important: check a or b; check c if applicable)	a. ☐ An entire facility	b. ☐ Part of a facility	c. ☐ A Federal facility
4.3	**Technical Contact Name**		Telephone Number (include area code)	
4.4	**Public Contact Name**		Telephone Number (include area code)	

4.5	**SIC Code(s) (4 digits)**	a.	b.	c.	d.	e.	f.

4.6	**Latitude**	Degrees	Minutes	Seconds	**Longitude**	Degrees	Minutes	Seconds

4.7	**Dun & Bradstreet Number(s) (9 digits)**	4.8	**EPA Identification Number(s) (RCRA I.D. No.) (12 characters)**	4.9	**Facility NPDES Permit Number(s) (9 characters)**	4.10	**Underground Injection Well Code (UIC) I.D. Number(s) (12 digits)**
a.		a.		a.		a.	
b.		b.		b.		b.	

SECTION 5. PARENT COMPANY INFORMATION

5.1	Name of Parent Company	☐ NA	
5.2	Parent Company's Dun & Bradstreet Number	☐ NA	(9 digits)

EPA FORM R PART II. CHEMICAL - SPECIFIC INFORMATION	TRI FACILITY ID NUMBER
	Toxic Chemical, Category, or Generic Name

SECTION 1.TOXIC CHEMICAL IDENTITY (Important: DO NOT complete this section if you completed Section 2 below.)

1.1	CAS NUMBER (IMPORTANT: Enter only one number exactly as it appears on the Section 313 list. Enter category code if reporting a chemical category.)
1.2	Toxic Chemical or Chemical Category Name (Important: Enter only one name exactly as it appears on the Section 313 list.)
1.3	Generic Chemical Name (Important: Complete only if Part I, Section 2.1 is checked "yes". Generic name must be structurally descriptive.)

SECTION 2. MIXTURE COMPONENT IDENTITY (Important: DO NOT complete this section if you complete Section 1 above.)

2.1	Generic Chemical Name Provided by Supplier (Important: Maximum of 70 characters, including numbers, letters, spaces, and punctuation.)

SECTION 3. ACTIVITIES AND USES OF THE TOXIC CHEMICAL AT THE FACILITY (Important: Check all that apply.)

3.1 Manufacture the toxic chemical:	3.2 Process the toxic chemical:	3.3 Otherwise use the toxic chemical:
a. ☐ Produce b. ☐ Import If produce or import: c. ☐ For on-site use/processing d. ☐ For sale/distribution e. ☐ As a byproduct f. ☐ As an impurity	a. ☐ As a reactant b. ☐ As a formulation component c. ☐ As an article component d. ☐ Repackaging	a. ☐ As a chemical processing aid b. ☐ As a manufacturing aid c. ☐ Ancillary or other use

SECTION 4. MAXIMUM AMOUNT OF THE TOXIC CHEMICAL ON-SITE AT ANY TIME DURING THE CALENDAR YEAR

4.1	☐ (Enter two-digit code from instruction package.)	

SECTION 5. QUANTITY OF THE TOXIC CHEMICAL ENTERING EACH ENVIRONMENTAL MEDIUM

			A. Total Release (pounds/year)(enter range from instructions or estimate)	B. Basis of estimate (enter code)	C. % From Stormwater
5.1	Fugitive or non-point air emissions	NA ☐			
5.2	Stack or point air emissions	NA ☐			
5.3	Discharges to receiving streams or water bodies (enter one name per box)				
	Stream or Water Body Name				
5.3.1					
5.3.2					
5.3.3					
5.4.1	Underground Injection on-site to Class I Wells	NA ☐			
5.4.2	Underground Injection on-site to Class II-V Wells	NA ☐			

If additional pages of Part II, Section 5.3 are attached, indicate the total number of pages in this box ☐ and indicate which Part II, Section 5.3 page this is, here ☐ (example: 1,2,3, etc.)

EPA Form 9350-1 (Rev. 04/97) - Previous editions are obsolete.

Range Codes: A = 1 - 10 pounds; B = 11 - 499 pounds; C = 500 - 999 pounds.

EPA FORM R PART II. CHEMICAL-SPECIFIC INFORMATION (CONTINUED)	TRI FACILITY ID NUMBER Toxic Chemical, Category, or Generic Name

SECTION 5. QUANTITY OF THE TOXIC CHEMICAL ENTERING EACH ENVIRONMENTAL MEDIUM

		NA	A. **Total Release** (pounds/year) (enter range code from instructions or estimate)	B. **Basis of Estimate** (enter code)
5.5	Disposal to land on-site			
5.5.1A	RCRA Subtitle C landfills	☐		
5.5.1B	Other landfills	☐		
5.5.2	Land treatment/application farming	☐		
5.5.3	Surface impoundment	☐		
5.5.4	Other disposal	☐		

SECTION 6. TRANSFERS OF THE TOXIC CHEMICAL IN WASTES TO OFF-SITE LOCATIONS

6.1 DISCHARGES TO PUBLICLY OWNED TREATMENT WORKS (POTWs)

6.1.A. Total Quantity Transferred to POTWs and Basis of Estimate

6.1.A.1. Total Transfers (pounds/year) (enter range code or estimate)	**6.1.A.2 Basis of Estimate** (enter code)

6.1.B. ___ POTW Name

POTW Address

City		State		County		Zip	

6.1.B. ___ POTW Name

POTW Address

City		State		County		Zip	

If additional pages of Part II, Section 6.1 are attached, indicate the total number of pages in this box ☐ and indicate which Part II, Section 6.1 page this is here ☐ (example: 1,2,3, etc.)

SECTION 6.2 TRANSFERS TO OTHER OFF-SITE LOCATIONS

6.2 ___OFF-SITE EPA IDENTIFICATION NUMBER (RCRA ID NO.)

Off-Site Location Name

Off-Site Address

City		State		County		Zip	

Is location under control of reporting facility or parent company? ☐ **Yes** ☐ **No**

Range Codes: A = 1 - 10 pounds; B = 11 - 499 pounds; C = 500 - 999 pounds.

EPA FORM R
PART II. CHEMICAL-SPECIFIC INFORMATION (CONTINUED)

TRI FACILITY ID NUMBER

Toxic Chemical, Category, or Generic Name

SECTION 6. 2 TRANSFERS TO OTHER OFF-SITE LOCATIONS (continued)

A. Total Transfers (pounds/year) (enter range code or estimate)	B. Basis of Estimate (enter code)	C. Type of Waste Treatment/Disposal/ Recycling/Energy Recovery (enter code)
1.	1.	1.M
2.	2.	2.M
3.	3.	3.M
4.	4.	4.M

6.2 ___ OFF-SITE EPA IDENTIFICATION NUMBER (RCRA ID NO.)

Off-Site Location Name

Off-Site Address

City | State | County | Zip

Is location under control of reporting facility or parent company? ☐ **Yes** ☐ **No**

A. Total Transfers (pound/year) (enter range code or estimate)	B. Basis of Estimate (enter code)	C. Type of Waste Treatment/Disposal/ Recycling/Energy Recovery (enter code)
1.	1.	1.M
2.	2.	2.M
3.	3.	3.M
4.	4.	4.M

SECTION 7A. ON-SITE WASTE TREATMENT METHODS AND EFFICIENCY

☐ **Not Applicable (NA)** - **Check here if no on-site waste treatment is applied to any waste stream containing the toxic chemical or chemical category.**

a. General Waste Stream (enter code)	b. Waste Treatment Method(s) Sequence [enter 3-character code(s)]	c. Range of Influent Concentration	d. Waste Treatment Efficiency Estimate	e. Based on Operating Data?
7A.1a	**7A.1b** 1 ☐ 2 ☐ 3 ☐ 4 ☐ 5 ☐ 6 ☐ 7 ☐ 8 ☐	**7A.1c**	**7A.1d** %	**7A.1e** Yes ☐ No ☐
7A.2a	**7A.2b** 1 ☐ 2 ☐ 3 ☐ 4 ☐ 5 ☐ 6 ☐ 7 ☐ 8 ☐	**7A.2c**	**7A.2d** %	**7A.2e** Yes ☐ No ☐
7A.3a	**7A.3b** 1 ☐ 2 ☐ 3 ☐ 4 ☐ 5 ☐ 6 ☐ 7 ☐ 8 ☐	**7A.3c**	**7A.3d** %	**7A.3e** Yes ☐ No ☐
7A.4a	**7A.4b** 1 ☐ 2 ☐ 3 ☐ 4 ☐ 5 ☐ 6 ☐ 7 ☐ 8 ☐	**7A.4c**	**7A.4d** %	**7A.4e** Yes ☐ No ☐
7A.5a	**7A.5b** 1 ☐ 2 ☐ 3 ☐ 4 ☐ 5 ☐ 6 ☐ 7 ☐ 8 ☐	**7A.5c**	**7A.5d** %	**7A.5e** Yes ☐ No ☐

If additional pages of Part II, Sections 6.2/7A are attached, indicate the total number of pages in this box ☐ and indicate which Part II, Sections 6.2/7A page this is, here. ☐ (example: 1.2.3. etc.)

EPA Form 9350-1 (Rev. 04/97) - Previous editions are obsolete.

Range Codes: A= 1-10 pounds; B=11- 499 pounds; C= 500 - 999 pounds.

EPA FORM R PART II. CHEMICAL-SPECIFIC INFORMATION (CONTINUED)	TRI FACILITY ID NUMBER Toxic Chemical, Category, or Generic Name

SECTION 7B. ON-SITE ENERGY RECOVERY PROCESSES

☐ **Not Applicable (NA) - Check here if no on-site energy recovery is applied to any waste stream containing the toxic chemical or chemical category.**

Energy Recovery Methods [enter 3-character code (s)]

1	2	3	4

SECTION 7C. ON-SITE RECYCLING PROCESSES

☐ **Not applicable (NA) - Check here if no on-site recycling is applied to any waste stream containing the toxic chemical or chemical category.**

Recycling Methods [enter 3-character code(s)]

1	2	3	4	5
6	**7**	**8**	**9**	**10**

SECTION 8. SOURCE REDUCTION AND RECYCLING ACTIVITIES

	All quantity estimates can be reported using up to two significant figures.	Column A Prior Year (pounds/year)	Column B Current Reporting Year (pounds/year)	Column C Following Year (pounds/year)	Column D Second Following Year (pounds/year)
8.1	Quantity released*				
8.2	Quantity used for energy recovery on-site				
8.3	Quantity used for energy recovery off-site				
8.4	Quantity recycled on-site				
8.5	Quantity recycled off-site				
8.6	Quantity treated on-site				
8.7	Quantity treated off-site				
8.8	Quantity released to the environment as a result of remedial actions, catastrophic events, or one-time events not associated with production processes (pounds/year)				
8.9	Production ratio or activity index				

8.10	Did your facility engage in any source reduction activities for this chemical during the reporting year? If not, enter "NA" in Section 8.10.1 and answer Section 8.11.			
	Source Reduction Activities [enter code(s)]	Methods to Identify Activity (enter codes)		
8.10.1		a.	b.	c.
8.10.2		a.	b.	c.
8.10.3		a.	b.	c.
8.10.4		a.	b.	c.

		YES	NO
8.11	Is additional optional information on source reduction, recycling, or pollution control activities included with this report? (Check one box)	☐	☐

* Report releases pursuant to EPCRA Section 329(8) including "any spilling, leaking, pumping, pouring, emitting, emptying, discharging, injecting, escaping, leaching, dumping, or disposing into the environment." Do not include any quantity treated on-site or off-site.

EPA Form 9350 - 1 (Rev. 04/97) - Previous editions are obsolete.

APPENDIX J

THE CLEAN SOUND COOPERATIVE, INC: WHAT IS IT?*

Clean Sound Cooperative is a non-profit regional oil-spill response organization funded by its industry members—including major oil, oil transportation, and pipeline companies.

Clean Sound responds to an oil spill much as a fire department responds to a fire. In the event of a spill, its crews and equipment are prepared for immediate response, regardless of the location, time of day or night, or weather conditions.

Oil spill cleanup is a complex process involving hundreds of people, dozens of pieces of major water-and land-based equipment, and several organizations that are all part of the oil spill response community—including federal, state, local, and tribal government agencies, private contractors, and the Clean Sound Cooperative.

While the individual/company whose vessel or facility was involved in a spill—the ''Responsible Party''—is responsible for its cleanup, the nature of government agency oversight and/or management will be determined by the size and conditions of the spill.

Clean Sound is an experienced partner in the oil spill response community, coordinating and cooperating closely and regularly with

- U.S. Coast Guard
- Washington State Department of Ecology, Emergency Management Division and other agencies
- Private contractor cleanup response firms
- Other West Coast oil spill cleanup cooperatives, including the Marine Spill Response Corporation

*Reprinted by permission of Clean Sound Cooperative, Inc.

Clean Sound owns, maintains and operates:

- A fleet of specialized oil spill response cleanup vessels
- Several trailers filled with boom, sorbents, and other cleanup equipment
- A mobile spill management center and a dedicated communications system

Above all, Clean Sound is people:

- Clean Sound crew members—more than 37 full-time crew members stationed at strategic sites in Puget Sound and the Strait of Juan de Fuca
- Backup contractor crews—additional personnel available if needed
- Industry Response Teams—personnel from the Clean Sound member oil refineries and other companies, specially trained to stage equipment and operate it if required

Clean Sound is also a planning organization, participating in the development of local, regional and state cleanup plans, as well as federal and state-organized spill drill exercises.

Clean Sound has a more than 23-year record of thorough preparedness and proven performance in handling oil spill cleanup in the region. It was organized in 1971.

THE INCIDENT COMMAND SYSTEM

The Incident Command System (ICS) is the management structure by which oil spill cleanup is organized in Washington. It:

- Provides a consistent structure for response to emergency situations.
- Is based upon teamwork, coordination, and cooperation between all entities involved, or potentially involved, in a spill response.
- Involves a unified command structure including predesignated On-Scene Coordinators who are in charge of the spill response operation for one or more agencies. The Coast Guard, Department of Ecology, and the responsible party may all be involved in joint management.
- Is coordinated by an Incident Commander who directs all phases of the spill response. Depending upon the location and extent of the spill, this

individual may be from the Coast Guard, Department of Ecology, or the Environmental Protection Agency.

THE RESPONSIBLE PARTY'S (THE SPILLER'S) ROLE:

- By federal law, the person or company causing a spill is fully responsible for its cleanup. They must hire the contractors needed. Their efforts will be monitored closely by government agencies responsible for oversight—the U.S. Coast Guard and/or the State Department of Ecology.
- When the responsible party is a Clean Sound member, Clean Sound's own crews and resources respond immediately as the first line of action. At a refinery, Industry Response Teams may be first line.
- Specific responsibilities include:
  - Assessment of spill
  - Establishment of a spill command post
  - Documentation/identification of the type of oil spilled
  - Containment of the oil spilled and protection of the environment, with particular emphasis on sensitive areas
  - Input relative to cleanup priorities
  - Timely and effective cleanup
  - Disposal of oil and oily waste
  - Restoration of damaged environmental/natural resources
  - Communication with local, state, and national response agencies
  - Communication with the media
  - Conducting of damage assessment (optional)
  - Paying for damages
  - Taking steps to prevent reoccurrence of spills and taking corrective action
  - Engaging in wildlife collection and care, in conjunction with responsible state/local/federal agencies

FEDERAL AND STATE OVERSIGHT

The Federal Role: U.S. Coast Guard

In marine spills, the U.S. Coast Guard is the federal government's lead agency responsible for overseeing oil spill cleanup efforts. In Washington, the Coast Guard has primary responsibility for all coastal marine and Puget Sound waters. If the party responsible for the spill is not available or is not performing adequately, the Coast Guard can assume direct control of the cleanup operation.

The State's Role

- The Washington Department of Ecology
  Ecology is the lead state agency, working in partnership with the U.S. Coast Guard to oversee environmental pollution response and cleanup efforts. Ecology may assume either a monitoring or a response/cleanup role, depending on the circumstances. Ecology has its own spill response program, with experienced spill response staff members in each regional office. Similar to Clean Sound, Ecology has its own mobile command center that can be set up near the site of a spill. Ecology also conducts formal investigations of oil spills, issues penalties when appropriate, assesses and collects damages resulting from oil spills, and approves oil spill prevention and contingency plans for shoreside facilities.
- Washington State Office of Marine Safety (OMS)
  Created as part of the Oil Spill Prevention and Response Act in 1991, the Office of Marine Safety (OMS) works to reduce the risk of oil spills in Washington by promoting safe marine transportation and adherence to prevention planning. OMS oversees oil spill prevention programs focused on improving human performance as the most effective means to prevent oil spills. Programs include vessel screening and monitoring, educational and technical outreach, prevention planning for tank vessels, vessel contingency planning and a marine information system. OMS also coordinates with other state agencies and regional marine safety committees to maximize resource use on spill prevention programs and develop regional marine safety plans.

MARINE SPILL RESPONSE CORPORATION (MSRC)

In the case of a major spill that exceeds the capabilities of Clean Sound and local response contractors, another level of industry-sponsored support comes into play—the Marine Spill Response Corporation (MSRC). Operating since 1993, MSRC is a nationwide, independent, not-for-profit corporation funded by oil companies and other organizations involved in the transport or storage of oil. Five regional centers are located in coastal areas around the United States. The Northwest Regional Office, responsible for spill response in Washington and Oregon, is based in Everett, Washington. Two 208-foot MSRC oil spill response vessels are located in the region—one in Everett and one in Astoria, Oregon—along with boom, skimming equipment, pumps and other equipment, and operators needed to respond to a major oil spill.

PAYING FOR SPILL CLEANUP

Primary Responsibility (The spiller)

Washington State's 1991 Oil Spill Prevention Act increases financial responsibility requirements to ensure those handling or transporting oil have the financial resources to conduct cleanups and pay for environmental damages.

- Tank vessels must demonstrate a minimum level of $500 million of financial responsibility.
- Cargo and passenger vessels over 300 gross tons must demonstrate responsibility of $500,000, or $600 per gross ton.
- Onshore and offshore facilities must maintain financial responsibility as determined by the Department of Ecology.

Back-Up Responsibility (If spiller isn't known, can't pay, or won't pay)

- Federal Trust Fund
  Oil Spill Liability Trust Fund, administered by the U.S. Coast Guard, if the spill occurs in areas under the Coast Guard's jurisdiction.
- Oil Spill Response Account
  Oil Spill Response Account, administered by the Washington State Department of Ecology, is coordinated with the federal trust fund to cover a spill where expenses exceed $50,000. A tax of 5 cents per barrel on crude oil delivered at marine terminals in Washington is divided in two ways:
- 3 cents goes into an administration account.
- 2 cents goes into a response fund.

INDEX